CYBERWAR AND REVOLUTION

Cyberwar and Revolution

Digital Subterfuge in Global Capitalism

Nick Dyer-Witheford and Svitlana Matviyenko

University of Minnesota Press
Minneapolis
London

Published by the University of Minnesota Press
111 Third Avenue South, Suite 290
Minneapolis, MN 55401-2520
http://www.upress.umn.edu

ISBN 978-1-5179-0410-4 (hc)
ISBN 978-1-5179-0411-1 (pb)

A Cataloging-in-Publication record for this book is available from the Library of Congress.

Printed in the United States of America on acid-free paper

The University of Minnesota is an equal-opportunity educator and employer.

26 25 24 23 22 21 20 19   10 9 8 7 6 5 4 3 2 1

Contents

Introduction: You May Not Be Interested in Cyberwar...

GRIM EPIGRAM

War, already distributed around the world by invasions, terrorist attacks, and drone strikes, insurgency and counterinsurgency, civil strife and foreign intervention, is today widening and intensifying in a new form waged across digital networks: cyberwar. Within capitalist democracies, warnings from national security agencies about Kremlin hackers, Chinese digital espionage, and jihadi virtual recruiters, not to mention networked leakers and whistleblowers, challenging said democracies from within, have been mounting for years; alarm rose to a fever pitch around Russian interference in the 2016 U.S. election, then went yet higher with the Cambridge Analytica scandal, and may ascend even further, exposing new actors or directing public attention, again, toward the usual suspects. So it is that in an era when the maxims of Marxist masters have fallen into deep disrepute, one at least seems to have escaped the oblivion of capitalism's memory hole. "You may not be interested in war, but war is interested in you," an aphorism ascribed to Leon Trotsky—first commander of the Red Army, no less!—is today not only widely cited but often, by emendation or implication, given a new, high-tech, exhortatory gloss: "You may not be interested in cyberwar, but cyberwar is interested in you" (Hoffman 2013; Schrage 2013; Dunlap 2014; Yong-Soo and Aßmann 2016). Like so much about cyberwar, however, the phrase suffers an "attribution problem" (Rid and Buchanan 2015), that is to say, an uncertainty as to authorship. For Trotsky never actually said or wrote "you may not be interested in

1

war"; the words were assigned to him in a spy novel, Alan Furst's (1991) *Dark Star.*[1] Yet despite, or perhaps because of, this grim epigram's status as a piece of viral misinformation, a digitally circulated non-Trotskyism, this book takes "you may not be interested in cyberwar, but cyberwar is interested in you" as a point of departure. We do so not because there is insufficient interest in cyberwar—there is now no shortage of commentary on the topic—but rather because so little of it is written from a politically critical perspective, committed to contesting the logic of the social system that daily draws the world closer to catastrophe. It is from such a position that we ask three introductory questions derived from non-Trotsky's cryptic proposition. First, what is "cyberwar"? Second, what could it mean to say that "cyberwar" is "interested in you"? And third—the issue that would surely have informed Trotsky's original observation, had he actually made it—what is the relation, today, of cyberwar to capitalism, and to revolution?

WHAT IS CYBERWAR?

Cyberwar, a neologism that asserts war has left the armored train from which Trotsky directed revolutionary troops far, far behind, is a term that has abruptly risen in prominence in recent years but that possesses more than a quarter century of genealogy (Healey 2013; Rid 2016). As we discuss later, *cybernetics* originated in the American and British military research of the Second World War, setting a path for the development of computers and networks that continued throughout the Cold War. However, the contemporary use of *cyberwar,* with specific reference to attacks in and on digital networks, did not emerge until the 1980s. Fred Kaplan (2016) suggests that U.S. military concern about this possibility was sparked by President Reagan's viewing of the film *WarGames* (1983), about computer-gaming teenagers breaking into the networks of U.S. Strategic Air Command. This anxiety-inducing event purportedly set in motion the first of what would become a long series of invariably urgent reports about the vulnerabilities of the United States (and its foes) to digital attack, produced by competing defense agencies and departments, only to be shelved, then rediscovered and repeated by successive administrations.

The actual conjoining of *cyber* with *war* was, however, the work of popular culture, reflecting the rapid uptake of science fiction author William Gibson's (1984) *cyberspace* to designate the increasingly widespread experience of internet use. According to Thomas Rid (2016), *cyberwar* first appeared in the digital avant-garde magazine *Omni* in a 1987 article about giant military robots. It was taken up more seriously in a 1992 essay by Eric H. Arnett in the *Bulletin of the Atomic Scientists* that declared, "The leading military concept of the new era might be called cyberwar" and applied it to a range of computerized "autonomous weapons," including crewless tanks, cruise missiles, advanced air-defense missiles, and anti-missile satellites. This probably inspired the *Chicago Sun-Times* news report of the same year, titled "Cyberwar Debate," about an alleged dispute between "scientists and the military" as to "who should wage war, man or machine," which the *Oxford English Dictionary* records as the earliest usage of the phrase.

In U.S. policy discourse, an early public salvo on "cyberwar" was the 1993 report by John Arquilla and David Ronfeldt for the RAND Corporation, the U.S. Air Force think tank. Dramatically titled "Cyberwar Is Coming!," it acknowledged the immediate influence of the U.S. forces' lightning Operation Desert Storm victory over Iraq in the Gulf War, in which Arquilla had been a consultant serving General Schwarzkopf. However, it reached back further into history to Mongol cavalry and Nazi Blitzkrieg to emphasize the importance of communication to gather knowledge crucial to military operations and overwhelm an enemy. From these instances, it extrapolated dramatic consequences from the information revolution and the growing use of computers and networks, suggesting that in the future, various forms of irregular warfare aided by such technologies would outmatch more heavily armed conventional forces. The authors developed this thesis in a series of subsequent publications, seizing on examples from the Zapatista uprising of Mayan peasants to the increasing powers of drug cartels to conduct the "swarming" operations characteristic of what they variously termed *netwar* or *cyberwar* (Arquilla and Ronfeldt 1996, 1997, 2000). At the time, their hypothesis seemed futurist provocation riding the coattails of fashionable discussions of digital smart weapons, but over the next decades, events would make "Cyberwar Is Coming!" prophetic (Arquilla and Ronfeldt 1993).

These included, first and foremost, the 9/11 destruction of the World

Trade Center, in part organized online; the subsequent "war on terror"; striking instances of Chinese cyberespionage; and Russian networked attacks on Estonia and Georgia. In 2010, Richard Clarke, an advisor on nuclear strategy and terrorism to the Reagan and Bush administrations, caught the mounting anxiety in a book *Cyber War: The Next Threat to National Security and What to Do about It,* which included lurid scenarios of society-crippling cyberattacks on the U.S. "critical infrastructure," with airplanes dropping out of the sky, communication systems collapsing, and military forces blinded (Clarke 2010, 64–68). Since then, furious controversy has raged over the plausibility of such prediction, with skeptics proclaiming "cyberwar will not take place" and dismissing the "threat inflation" funding a new digital–military complex (Rid 2013; Blunden and Cheung 2014) and counter-counterblasts announcing with equal assurance that "there will be cyberwar" (Stiennon 2015) or, indeed, that "cyberwar is already upon us" (Arquilla 2012). Around these debates, a wave of cyberwar discourse has spread out, initially largely generated by military analysts, intelligence specialists, and political scientists (Miller and Kuehl 2009; Libicki 2009; Clarke 2010; Ventre 2011; Kerschischnig 2012; Healey 2013; Singer and Friedman 2014; Kaiser 2015; Valeriano and Maness 2015; Kaplan 2016; Klimburg 2017) but, in the last few years, reaching out to wider audiences through investigative journalism (Zetter 2014a; Harris 2014; Gertz 2017; Patrikarakos 2017; Sanger 2018) and "next war" technothrillers (Singer and Cole 2015). As we finalize this book, yet more works on what has become a "hot" topic are appearing.[2]

This cyberwar discourse is replete with definitional disputes. If we start just with the "cyber" component, it can be accepted that the weaponry of cyberwar is computer code, specifically code "used or designed . . . with the aim of threatening or causing physical, functional or mental harm to structures, systems or living beings" (Rid and McBurney 2012, 7). Cyberattacks, therefore, can include multiple forms of hacking, from denial of service attacks that paralyze an opponent's websites to interruptions of military network communication to the dreaded critical infrastructure–targeting malware that could potentially disrupt factories, electricity grids, transport networks, and even the command-and-control systems of nuclear arsenals (Gartzke and Lindsay 2017). Wider definitions, however, also sometimes encompass viral propaganda; online psychological and virtual political manipulation, including "fake news"; and strategic information leaks and electronic electoral tampering. This list could be extended to cover the surveillance and profiling systems that are deployed as defensive weapons against such activities. However, some Western experts prefer to address such psychological and political operations under wider headings, such as "information war" (Porche et al. 2013) or "netwar" (Arquilla and Ronfeldt 1996; NATO 2006); Russian and Chinese authors also adopt these wider formulations, which reflect differences in national strategic doctrines that we take up later in this book.

The issue of *cyber*war is further complicated because the "cyber" in cyberwar may be distinct from, preliminary to, or simultaneous with the "kinetic" use of jet bombers, helicopter gunships, artillery, rocket batteries, tanks, mortars, small arms, and other conventional weapons (Clarke 2010). Unlike many other theorists of cyberwarfare, we do not oppose "cyber" and "kinetic" weaponry but rather emphasize the ways in which they cross over and complement each other until it is difficult, if at all possible, to distinguish between them. This has been particularly apparent in recent conflicts like those in Ukraine and Syria that at once entail fierce battlefield fighting (itself involving electronic command, control, communication, and weapons targeting systems); social media mobilization of rival popular unrests; aggressive digital surveillance and entrapment of opponents by state regimes; and hacking of various kinds, both from within and outside the given country. Such conflicts are today often described as "hybrid" cyberconventional wars, in which the two elements either fully coincide or phase in and out during complex, asymmetrical military confrontations (Murray 2012).

The "war" in cyberwar is equally vexed. *Cyberwar* is a term most often applied to digital operations conducted by or on behalf of nation-states against geopolitical opponents. Some experts adopt a minimalist definition of cyberwar as an extension of the nation-state's legitimated monopoly on violence and may also stipulate that it must inflict a level of destruction comparable with a major use of conventional weapons to constitute an act of war. This approach effectively demands that acts of cyberwar must, to deserve the name, be recognizable within the terms set by Carl von Clausewitz's nineteenth-century classic of military theory *On War*. It thus strives to sharply distinguish cyberwar proper from mere cyberattacks, cybercrime, cyberespionage, and affine activities (Rid 2013; Healey 2013).

However, it is increasingly difficult to uphold this distinction. Actions against states by nonstate actors, such as hacker collectives like Anonymous or "lone hackers" like Guccifer, Guccifer 2.0, or Phineas Fisher, or by quasi-state actors, such as ISIS, and by states against such actors are also now often described as cyberwar. The question of when these activities enter the realm of "irregular war," "guerrilla war," "ghost war," or "stealth war" remains contested (Singer and Friedman 2014). Even more challenging to Clausewitz's theory is that state cyberwar activity is frequently undertaken via unofficial or semiofficial para-state proxies in a shadow world of privatized or criminal hacking (Bowden 2011; Harris 2014; Maurer 2018). In contemporary diplomatic and military thought, there is an increasing tendency to say that a serious cyberattack is not strictly comparable to the use of conventional weapons and more closely resembles a large-scale cybercrime and might be considered an act of war and invite retaliation as such (Knapton 2017; Crisp 2017). The 2018 U.S. review of nuclear weapons policy opens the door to their use in response to a massive cyberattack (Kaplan 2018).

These definitional debates are not abstract semantic disputes. As Robert Kaiser (2015) argues, since 2000, *cyberwar* has named an emergent "knowledge power assemblage," mobilizing institutional powers, lavish budgets, and lucrative career paths. It has become the core of a discursive "resonance machine," with its own "catalyzing agents," "shimmering points" of elaboration, refinement and subtle discrimination, and "centers of calculation," focused around the materialization of a "new policy object" (Kaiser 2015, 11–20). In 2009, after tortuous bureaucratic infighting between competing governmental departments and military agencies, the Pentagon created a Cyber Command to conduct operations in a "cyberdomain" now deemed as important as land, sea, air, and space. This is a logic being duplicated within national security states around the world. The observation that cyberwar has now become extremely handy for acquiring big funds for research and militarization (Rid 2013) is true, but, with a self-validating circularity, the cyberwar apparatuses built on this promotional discourse not only speak but also prepare and act.

Public knowledge of these cyberwar acts is, however, limited and distorted; if the new cyberwar complexes are in one way new "assemblages of knowledge and power," they are also assemblages of power, nonknowledge and antiknowledge, military factories of ignorance and obfuscation. War always involves secrets, but cyberwars are exceptionally deeply cloaked. As Michael Kenney (2015, 117) observes, "cyber-warfare is typically a covert form of statecraft. Herein lies much of [its] utility as a weapon: states can attack their adversaries without declaring war against them." If there is one cyberwar cliché more common than "you may not be interested in cyberwar," it is a redo of Clausewitz's "fog of war" as "fog of cyberwar" (Valeriano and Maness 2012; Canabarro and Borne 2013; Rantapelkonen and Salminen 2013). Digital war is a veritable fog machine, because operations are usually conducted covertly and are often intended to confuse; hacks are hidden from view and, when discovered, are laden with misdirection; signals intelligence implodes into infoglut. Apparently incriminating traces of hacking tools left at the site of a digital breach may be intentional false flags and what looks like a singular exploit actually a chain of attacks by different actors "piling on" to a discovered system vulnerability. Is a hacking exploit that leaves as a signature the name of the founder of the Soviet secret service a careless, or cavalier, disclosure of national responsibility—or a blatant red herring? As of today, it can be said that all, or nearly all, obvious indicators pointing to a potential responsible party, such as time, specifics of code, politics or other motifs, and cybertactics, can be manufactured to point in a wrong direction. In investigating cyberwar, conspiracy theory is methodologically mandatory. However, not only attribution of attacks but also the distinction between attacks and accidents becomes highly problematic; "fog of war" (ignorance) and "friction of war" (mishap) intermingle. Cyberwar thus tends to "obliterating proof" (Filiol 2011, 260), sometimes by producing too many proofs, "incessant fluctuations" of proofs, to use Claude Shannon's phrase for noise, or, in the words of W. Ross Ashby ([1963] 1966), "variety [that] can destroy variety."

Cybersecurity experts argue that, given sufficient time, it is indeed possible to make a reasonable determination of responsibility for cyber-attacks. This process can involve analyzing technical details, such as the design of malware payload or "IP addresses, email addresses, domain names, and small pieces of text" used in the attack; comparing the modus operandi of hackers with the known profiles gathered from other cases; and assessing the likely reasons for an intrusion—but even then, verification is a matter of degree of certainty rather than a binary yes or no answer (Rid and Buchanan 2016). In this book, we often necessarily rely on such "balance of probability" interpretations, but with an awareness that cyberwar is a weaponization of information that always threatens to destroy truth. It is, as Justin Joque (2018, "Foreword") writes, a force of "machinic deconstruction" that constantly threatens "to sweep away any belief in security." We write also knowing that, in "a world where it is not clear who is hacking whom" (Ridgeway 2017), the attribution of cyberwar activity can itself become a tactic setting up an adversary for sanctions and retaliation. Academic interpretation, too, can be weaponized; those who stare into the abyss of cyberwar soon find it staring back at them.

Yet even the dense smog of cyberwar today suggests underlying fires of smoldering, sometimes blazing, network conflicts. These conflicts are indexed in the "science fiction poetry" of code names assigned by military and security agencies to specific cyberoperations (Shilling 2017). Thus, in the lexicon of U.S. intelligence, "Operation Olympic Games" designates the Stuxnet computer worm, now generally held to be a joint project of U.S. and Israeli agencies between 2006 and 2012 to disable more than a thousand centrifuges at Iran's Natanz nuclear research facility—a "Rubicon-crossing" moment of state-sponsored digital aggression (Zetter 2014a). Its planned successor, Operating Nitro Zeus, suggesting "a Greek god on steroids" (Shilling 2017), was a more comprehensive cyberattack plan to disrupt and degrade communications, power grid, and other vital systems had Iran not suspended its nuclear weapons program. Smaller scale, but of the same nuclear preemptive ilk, was Operation Orchard, an Israeli hybrid operation in which a 2007 air strike against a suspected nuclear weapons factory in Syria was supported by the digital disabling of that nation's antiaircraft defenses. And in 2017, continuing the apparent tradition of classical naming for cyberoperations, the U.S. military reported on the actions of Joint Task Force Ares, charged with destroying and disrupting the computer networks of ISIS, an operation so highly classified it could not be described other than by the assertion that it "provided devastating effects on the adversary" (Lamonthe 2017).

Many Western code names document epic battles with clandestine cyberopponents. Operation Titan Rain was an alleged cyberespionage attempt, an "advanced persistent threat" (APT) ongoing from 2003, by hackers from the People's Republic of China—possibly members of the People's Liberation Army—to penetrate the networks of U.S. defense institutions, military contractors, and high-technology businesses. Operations Shady Rat, Aurora, and Night Dragon identify a later round of similar attacks, commencing in 2006. Other names catalog the digital intrusions against American institutions credited to agents of first the Soviet Union and later the Russian Federation. Operation Moonlight Maze, an FBI inquiry into a series of probes into the networks of the Pentagon, NASA, the Department of Energy, private universities, and research labs that started in 1998, and its successors, Operation Makers Mark and Operation Storm Cloud, are sometimes called the first major cyberattacks in history; in Operation Buckshot Yankee, in 2008, a malicious code placed on a flash drive uploaded itself to networks of the U.S. military's Central Command, also attributed to Russian agents; and, most recently, the activities of the hacker groups Cozy Bear and Fancy Bear in Operation Grizzly Steppe leaking the communications of the Democratic National Committee in the 2016 elections.

Other events not awarded exotic military cryptonyms nevertheless add to the archive: Russian hacking attacks on Estonia in 2007 and Georgia in 2008; the 2012 hacks using the Shamoon computer virus against the national oil companies of Saudi Arabia's Saudi Aramco and Qatar's RasGas, considered an Iranian retaliation for the Stuxnet attack; the 2014 disruption of Sony Pictures computer networks, widely, but still controversially, considered the work of North Korean agents against the portrayal of Kim Jong Un in the satirical film *The Interview*; the subsequent brief but total blackout of North Korea's internet, suspected as the work of either U.S. state agencies or freelance American hackers; the 2015 cyberattack on Ukraine's electrical system, using malware known as BlackEnergy and launched from Russian IP addresses, that left 230,000 people without electricity for a period from one to six hours, considered to be the first known successful cyberattack on a power grid; the occult conflicts of the Equation Group and Shadow Brokers, which resulted in the theft of secret U.S. cyberwar weapons and their subsequent use in criminal and probably political ransomware attacks, such as the WannaCry and Petya exploits of 2017. And this catalog omits the nameless, everyday computer attacks, digital spying, viral propaganda, and other virtual mayhem perpetrated on and beyond the hybrid battlefields, such as those of Donbas, Kurdistan, or Yemen, whose ruins are perhaps the starkest reminder of what is ultimately at stake in in the rise of cyberwar.

INTERESTED IN YOU?

If Trotsky *had* said "you may not be interested in war, but war is interested in you," he might have been doing no more than reiterating a classic Marxist trope about the inescapable involvement of individual lives in the vast dialectic of historical forces. Yet the idea of a hypostatized "war," an impersonal apparatus that assumes a self-propelling autonomy, even while being interested in you, the individual subject, also conveys a blend of awful automatism and dread intimacy. If war is "interested" in you, it (to rehearse the *Oxford English Dictionary* [*OED*] entry) "attends to you, is both curious and concerned about you," and "seeks its own advantage through you," even as it presumably prepares you to kill, or be killed, or both. It is "into you," *really* into you. This military "interest" in "you"— the moment in which war engages, infiltrates, excites, recruits, suspects, mutilates, and annihilates the subject—is a crucial aspect of cyberconflicts.

Warfare permeated the entire fabric of everyday life in the "total wars" of the twentieth century, with their mobilization not only of vast armed forces but of the industry, hospitals, bureaucracies, logistic systems, and psychological and propaganda apparatuses that supported the battlefronts (Black 2006; Dreiziger 2006; Chickering, Förster, and Greiner 2005). The persistent power of national security apparatuses in times of Cold War, memorialized in Eisenhower's reference to the "military–industrial complex" and continuing to the present, has been well documented (Brooks

2016). Yet it might seem that, with nuclear dangers supposedly pacified since 1989 by the advent of global capitalism, and the deployment of conscripted armies largely replaced by highly professionalized special-forces operations, an all-encompassing military activation of society is past. The recognition that digital technologies might raise the reach of war to a new scope and intensity has been slowly dawning.

What this reach grasps is the "datified subject" (Cheney-Lippold 2017, 35), that is to say, the identity constituted for individuals by their activity on digital networks. This is not to suggest an entire virtual translation of the person into the network but rather the collection of digital identifiers that converge on, add to, overlay, interpenetrate, and, in some contexts, supplant the multiple social codes and indicators of which corporeal individuality is composed. There is now an extensive literature on the various corporate and state "profiling machines" (Elmer 2004) that assemble and ascribe identities from the traces of digital searching, liking, browsing, shopping, chatting, and navigating, and, indeed, from the mere act of carrying location-transmitting mobile phones. And there is a similar series of investigations into the "black boxes" (Pasquale 2015; see also Amoore 2013; Finn 2017) of algorithmic processes that, from the accumulation of "big data," predict future activities—and of the many biases and errors that enter this process (O'Neill 2016). This literature highlights the paradox that the mass volumes of big data enable increasingly precise, personalized mapping of an individualized "data subject" (Bauman et al. 2014) according to its conformity or deviation from a series of patterns. We will follow this idea through three different registers in which war touches the "datified subject": battlefield intelligence, civilian surveillance, and digital propaganda, focusing particularly on examples from the United States.

Shawn Powers and Michael Jablonski (2015, 81) observe that origins of data profiling lie in intelligence operations, where *mosaic theory,* the idea that "one can piece discrete bits of information together to predict the likely intentions or actions of others," has been a core idea. Early in the so-called war on terror, mosaic theory was applied at the CIA's black sites and U.S. military's detention centers at Guantánamo, Bagram, and Kandahar, where captives were often held and interrogated solely on the basis of association with other suspects (Pozen 2005). The idea that big data can generate an "individualization" of war has become increasingly prominent in cyberwar doctrine. In 2014, Charles Dunlap, a retired U.S. major general (and one of the several authors who invoke the pseudo-Trotsky aphorism we started with), suggested that anticipations of cyberwar as a series of "apocalyptic" infrastructure attacks might be less important than the "hyperpersonalization of war" based on data profiles. The possibility that particularly excited him was that of targeting specific members of an opponent's forces, "primarily the leadership cadre but also critical technicians and experts," with the primary example being the use of drones, both for information gathering and attack delivery from the skies over Afghanistan, Iraq, Pakistan, Yemen, and Somalia.

A year later, a report for the U.S. Army War College Strategic Studies Institute by a serving officer, Colonel Glenn Voelz (2015), *The Rise of iWar: Identity, Information, and the Individualization of Modern Warfare*, developed a similar proposition more broadly. Referring again primarily to U.S. counterinsurgency operations, Voelz describes how war against "wayfarer warriors, eager to apply their skills and experience across multiple theaters of conflict and acts of terrorism" in a "persistent global circulation of fighters," has presented an "entirely new security dilemma for nation-states" (14). In response, he suggests, U.S. forces are bringing "new tools to the battlefield specifically designed to support identity management and personality-based targeting" (10). These include drones; biometric tool kits capable of "13 point identification (10 fingers, two irises, and one face)"; and "expeditionary forensics" using advanced data processing systems and effecting a "fusion of identity attributes," including "biologic, biographic, behavioral, and reputational information" (22). Voelz presents these developments as the fruition of previous generations of U.S. military thought on cybernetic warfare. Now, however, it amounts to a "new paradigm of war" differing from the "mass based depersonalization" of military operations under the "Westphalian state system." This paradigm is capable of systematic disaggregation of threats down to the lowest possible level, often the "individual combatant on the battlefield," based on three elements: "individualization, identity, and information" (44).

As John Cheney-Lippold (2017, 37) makes clear in his *We Are Data*, a drone attack vaporizing targets and bystanders with Hellfire missiles can be delivered on the basis of digital metadata,

> data about data . . . data about where you are, from where you sent a text message, and to where that message is sent . . . data that identifies what time and day you sent that email, and even the type of device you used to send it.

This is the perhaps the clearest demonstration of the fateful intersection between data subject and the corporeal body. Later in this book, we will return to cybernetic counterinsurgency doctrines and question their claims of clinical accuracy and that they supplant mass war; on the contrary, we will suggest, cyberwar bridges small-scale "dirty wars" and the species-scale threat of nuclear war. But for the moment, we will leave the battlefield to consider how "iWar" (Gertz 2017) manifests an "interest" in its subjects' civilian existences through two of its other aspects: surveillance and propaganda.

When, in 2013, Edward Snowden fled to Hong Kong and turned over to journalists of the *Washington Post* and the *Guardian* internal documents disclosing the scope of U.S. National Security Agency surveillance programs, he revealed cyberwar as a project of vast, almost phantasmagoric, scale. As one journalist observed, "it . . . sounded like fiction from some deranged person wearing a tinfoil hat. But it was true" (Kravets 2016). Of this massive cache, whose contents are still only gradually being processed, the aspect that has attracted most attention from U.S. citizens is the scope of their government's surveillance programs initiated in the wake of the 9/11 attacks on the World Trade Center. Perhaps the greatest shock of the Snowden Files came from the discovery of a National Security Agency (NSA) bulk collection of metadata relating to U.S. telephone calls to and from foreign sources. But almost equal alarm was aroused by the PRISM program, enabled under President Bush by the Protect America Act of 2007 and by the Foreign Intelligence Surveillance Act Amendments Act of 2008, whose monitoring was conducted with the compliance of corporations including Microsoft, Yahoo!, Google, Facebook, YouTube, AOL, Skype, and Apple. By 2011, the NSA and other U.S. security agencies were, with the cooperation of the private sector, "collecting and archiving about

15 billion Internet metadata records every single day" (Deibert 2015, 11). U.S. citizens discovered that their government's security apparatus was "interested" in them to a degree such that it monitored the addresses, locations, times, and patterns of their online communication on devices they regarded as personalized prosthesis and networks sold to them as constitutive of social life and friendship—a cultural trauma, even if it was one that (like all trauma) subsequently became normalized.

What this level of "interest" may mean is made apparent by later revelations as to what inclusion on U.S. watchlists of terror suspects means (Scahill 2016). Such watchlists "gather terrorism information from the sensitive military and intelligence sources around the world" (Scahill 2016, 18), including social media. Inclusion on such a list, particularly the comprehensive Terrorist Identities DataMart Environment (TIDE), is a fateful event, because "once the US government secretly labels you as a terrorist or terrorist suspect" and distributes this information to airlines, police, "private entities," and other institutions, "it can become difficult to get a job or simply to stay out of jail. It can become burdensome—or impossible—to travel. And routine encounters with law enforcement can turn into ordeals" (Scahill 2016, 18). It was estimated in 2014 that about 1 million names were on such lists. The process by which people are put on watchlists is opaque, and there is no clear mechanism for appeal. Therefore, for those mistakenly added, the questions, Am I on the watchlist? How did I get on the watchlist? Can I get off the watchlist? become a desperate accompaniment to cyberwar's personal interest in the "datified" or "data subject."

It might seem tendentious to include surveillance of domestic populations as part of cyberwar. In fact, the NSA programs revealed by Snowden also included a major component of surveillance and penetration of computers worldwide, a point we return to later in this book. But even more to the point is that the origin of PRISM, TIDE, and other major U.S. surveillance programs is the "war on terror" and the fear that the "wayfarer warriors" of jihad would find their way to, or spring up indigenously within, the homeland. It is precisely fear of the networked mobilization of domestic populations by foreign, external, or alien powers—the state's dread that its citizens might become subjects of interest to other powers—

that is a feature of cyberwar. It was this anxiety that would, three years after Snowden's disclosures, give rise to the second great North American cyberwar trauma: the Russian "hacking" of the 2016 presidential election.

Investigative autopsies are still being performed on this event; it is the contentious uncertainty as to what actually happened, a paradigmatic case of the "attribution problem" described earlier, that makes its political impact so chaotic. Nonetheless, we can attempt a provisional narrative. The scandal began during the election with the leaking of politically embarrassing emails between members of Hillary Clinton's campaign team and the Democratic National Convention (DNC). These leaks are now widely, though not universally, held to have been the work of two hacker groups, agents of or proxies for the Russian government, known in cybersecurity circles as APT 28 and 29, or by the code names Cozy Bear and Fancy Bear, seeking to damage the Democratic campaign and assist in the election of Donald Trump. The issue reportedly caused President Obama to threaten President Putin, via the "red phone" hotline of Cold War atomic crisis fame, with digital retaliation (Arkin, Dilanian, and McFadden 2016).

While the issue of the DNC leak was still under investigation, however, the issue widened with allegations that Russian operatives had also attempted to sway the election by generating content on social media, such as Facebook and Twitter. These allegations, first treated with incredulity by the owners of the platforms, have subsequently been verified, with the discovery that a number of relatively inexpensive but nevertheless widely circulated, fabricated news stories and falsely attributed election advertisements were purchased by the Internet Research Agency (IRA), a St. Petersburg company well known for its online influence operations on behalf of the Russian government domestically, in Ukraine, and in the Middle East.

Once again, the U.S. population discovered that social media and mobile phones, technologies they were trained to treat as intimate extensions of the self, were portals for the state. Now, however, the intruder was not their own government but a foreign power. The outrage at this violation of liberal democracy certainly displays naivety and disingenuity; the U.S. national security state has been a famous perpetrator of election interference abroad (O'Rourke 2016; Emridge 2017; Gill 2017). But the scandal raised profound questions about digital media. Not only did the chronic insecurities of email permit the DNC hack but it was the news feeds and filter bubbles, posts and retweets of social media, that allowed such disclosures to be incorporated into a broader propaganda campaign. In this sense, the populations of platforms such as Facebook and Twitter, dependent on "user-generated content," appear not just as the passive victims of misinformation but its complicit abettors. A nation of social media users were suddenly cast, not as empowered and emancipated networkers, but as "useful idiots."[3] At the end of 2017, Facebook, in an attempt to counter the widespread criticism it had received for its uncurated and unedited circulation of "fake news," made a gesture exemplary of the individualization of cyberwar's "interest" in its "datified subject," launching a digital tool to allow users to see if they personally had liked or followed "Russian propaganda accounts" (Levin 2017).

In its wake, this scandal (which we discuss further in chapter 2) has generated a new concept, *computational propaganda* (Bolsover and Howard 2017; Bradshaw and Howard 2017), referring to state-organized campaigns of digital persuasion against political adversaries. Unsurprisingly, investigators have found that unfortunate events shocking in the imperial heartland were common elsewhere around the planet, tracking incidents of computational propaganda in some thirty countries. Campaigns are characteristically conducted across social media platforms by some permutation of troops, trolls, and bots. That is to say, they involve humans, who may be direct employees (such as soldiers or public employees), paid proxies, private contractors, or patriotic volunteers, very often working with automated algorithmic software agents capable of message dissemination and rudimentary online conversation. Such campaigns may engage issues either in terms of *affirmation*, advancing a particular idea or cause; *negation*, criticizing, abusing, and threatening those who hold a contrary position; or *distraction*, deflecting conversations away from particular issues or disrupting them with spam. Operatives often operate under aliases and frequently generate deceptive and provocative content, especially false news stories, supported with doctored photos and videos. Campaigns may be cast broadly or narrowly targeted; trolling efforts are often individualized, harassing specific political actors, journalists, or bloggers.

Just as propaganda per se may or may not be immediately related to war, so too computational propaganda does not necessarily have a direct military connection. Much of such activity is aimed at controlling domestic populations, where it goes hand in hand with surveillance. There is also, as the U.S. election demonstrated, a broad similarity between political-party digital campaigning and state-sponsored computational propaganda. At the same time, computational propaganda campaigns can assume military dimensions when they are conducted in support either of conventional military operations or in combination with other disruptive cyberattacks— or, arguably, in an attempt to elect politicians whose military and foreign policies might be favorable to the perpetrator, or simply to disturb and demoralize an opponent. The United States and NATO have accused the Russian state of such campaigns in Estonia, Georgia, and Ukraine and in elections in both the United States and Europe. The United States itself engages in computational propaganda campaigns; a U.S. military plan created "sock puppet" social media accounts to influence Middle Eastern public opinion (Fielding and Cobain 2011), and, as we discuss later, in Russia's view, U.S. State Department support for "color revolutions" in Eastern and Central Europe is regarded as a form of "information war."

In cyberwar, hyperpersonalized tactics destroy the enemies of the state, ubiquitous surveillance monitors its suspects, and "computational propaganda" mobilizes it supporters, at home and abroad. Battlefield intelligence, surveillance systems, and propaganda campaigns are all long-standing manifestations of war's "interest" in the populations it activates and assails. What is new about the cyberwar is that access to the "datified subject" both expands the scope of such operations across planet-spanning networks and intensifies the precision with which they can be targeted. War's "interest" in "you" and "I" assumes a new cognitive and affective intimacy via technological systems that have become "extensions of our nervous system" (McLuhan 1994), to which we have outsourced the maintenance of social ties, memories, and collective knowledge. Paradoxically, this personalization of war depends on an automated apparatus of completely impersonal scale and speed: algorithmically processed big data from which drones are targeted, watchlists drawn, and social media bots assigned their missions. Cyberwar is not (yet) as inhuman as other forms of war in terms of dead and wounded but is to an unprecedented degree *ahuman* war, run by malware implantation, bot nets, chatbots, the tweaking of algorithmic filter bubbles, big data scanning surveillance, and the "kill chains" of drone warfare (Deibert 2013; Cockburn 2015; Shaw 2016). As such, it can be seen as part of the tendency of capital to automation and the expansion of constant capital, such that humans occupy increasingly interstitial and marginal positions within destructive, as well as productive, systems. These are transformations that, we will suggest, deeply alter not only the military but the ideological relations of states to their subjects; but before we turn in that direction, we should place the rise of cyberwar in a larger political context.

CYBERWAR—AND REVOLUTION?

Trotsky, we speculate, would be appalled at the historical moment in which "you may not be interested in cyberwar" has gained currency. For the time of cyberwar is the time of the global triumph of capitalism. The predictions of the revolutionary who, in 1932, declared "capitalism has outlived itself as a world system" seem today utterly confounded. Only the grimmest of vindications could be drawn from the fulfillment of his forecasts about the Stalinist betrayal of revolution, the degeneration of socialism into bureaucratic despotism, and its possible eventual reabsorption by the world market (Trotsky [1937] 1973)—and not just in the former USSR but worldwide, with the the People's Republic of China's compromise with capitalism providing another, albeit differently nuanced, iteration of the same story.

This is not the place for a detailed examination of the political economy of "postsocialist" nations (Lane 2014). In the USSR, the orgy of privatization and the chaos of shock-therapy marketization after 1991 have been followed under Putin by a new period of nationalization, in which oligarchic capitalists operate under supervision of, and overlap with, state elites in a context of market exchange (Worth 2005; Pirani 2010; Dzarasov 2014; Sakwa 2014). In the more complex case of the People's Republic of

China, the party apparatus has maintained control of some "commanding heights" of the economy, even as other domains are handed over to private ownership, including both foreign and domestic investors. Some commentators see China's arrangements as tantamount to full subsumption with the world market (Li 2009; Hart-Landsberg 2013); others consider that a "left turn" from within the party apparatus could reestablish a socialist project (Amin 2013). At the moment, the latter possibility seems remote. In Russia, China, and other postsocialist countries, the wage organizes production, the responsibilities of the state for provision of public services have typically been diminished, and there are vast income gulfs between workers in and owners of the means of production. Corruption is often rife. Political elites are either identical with, entangled in, or dependent on capital ownership, and although these elites must in various ways manage, mobilize, and hegemonize public opinion, their fractions define the agenda of policy decisions, and perhaps especially foreign policy decisions. Thus discussions of international relations in terms of the intentions or desires of "China" or "Russia"—as, of course, of the "United States," "Canada," "Ukraine," "Saudi Arabia," or any other capitalist nation—must be understood as shorthand metonymic mystification of ruling-class power.

Cyberwar has therefore emerged as a topic of global concern at a moment when the teleological certainties of Marxism seem broken or reduced to cruel caricature. This is not a coincidence. As we will argue, the emergence of cybernetics from the military–industrial complex of the United States at the end of the Second Word War was an important part of that nation's ascent as a new imperial leader for the capitalist system. Computers and networks, both in their military and economic applications, played an important role in eventual U.S. victory over the USSR in the Cold War. And their extension into electronic commodities, industrial automation, supply-chain logistics, and financial trading was a crucial part of the globalization in which a reinvigorated capitalism from 1989 on disseminated itself around the planet, under the shelter of the global hegemon's cruise missiles, smart weapons, and satellite intelligence. This armed pacification of a world market has, however, not had the finality many expected. Rather, it has generated new wars, of two major types, both misnamed and ill defined but each a consequence of capital's global triumph over its socialist opponents.

The "war on terror" is, of course, the conventional and ideologically laden name for the protracted sequence of conflicts set in motion when mujahideen, armed and financed by the United States and its Saudi Arabian ally to fight the Soviet Union in Afghanistan, turned on its imperial patron with the destruction of the World Trade Center in 2001. These conflicts, centered on the Middle East but radiating across the planet, include the invasion and occupation of Afghanistan; world-distributed terrorist attacks; and counterterrorist operations across Pakistan, Yemen, Somalia, Nigeria, the Philippines, Mali, Libya, and many other theaters. If the "war on terror" is sometimes colloquially used to include the U.S. invasion of Iraq, this craven acceptance of the spurious rationalizations offered by the Bush administration could be only retroactively justified, as U.S. occupation generated first both Sunni and Shia insurgency and later, in its aftermath, the rise of ISIS. This so-called war on terror interpenetrates other regional conflicts, such as those in Kashmir between India and Pakistan, Russian actions in Chechnya, and Saudi Arabia's intensifying clashes with Iran in Yemen and elsewhere, and also overlaps with Israel's constant operations against Palestinians, wars with neighbors, and determination to maintain its regional monopoly of nuclear weapons. It is not our aim here to map the noxious vectors of the "war on terror," only to highlight how its mutating fronts have been a bleeding edge for the development and use of cyberweaponry, in counterinsurgency operations, domestic surveillance, and digital strikes and sabotage against nuclear weapons facilities.

The other kind of war breaking out since 1989 is the so-called New Cold War (Sterling 2007; Moss 2013; Jones 2016) between the capitalist "winners" and postsocialist "losers" in the previous round of conflict. The phrase is most often used with reference to deteriorating U.S. relations with Russia around issues such as U.S. intervention in the disintegration of Yugoslavia; NATO expansion into post-Soviet republics in Central and Eastern Europe; populist "color revolutions" in Ukraine, Georgia, Kyrgyzstan, Belarus, and Moldova; and Russia's response to such changes of regime. "New Cold War" is also, however, applied to similarly deteriorating U.S. relations with China, specifically focused around trade relations, military control of the South China Sea, and, more recently, expanding Chinese influence in Europe, but more generally attributable to U.S. apprehensions about the rising power widely expected soon to outstrip it as the world's largest national economy.

As several critics have observed, applying the Cold War metaphor to these tensions is deceptive (Ciuta and Klinke 2010; Sakwa 2013; Budraitskis 2014). At stake in the old Cold War were contending modes of production and opposed models of society, organized respectively around markets and planning—a compelling binary whose hold over both popular and elite imaginations was scarcely diminished by the actual corruption or occasional collusion of the antagonists. In the New Cold War, no such epic binaries are in play; all protagonists are participants in the world market, differentiated at best along a spectrum that runs from free-market neoliberalism to variants of state capitalism. The one exception to this pattern, an exception that proves the rule, is the confrontation between the United States and North Korea, a hybrid of failed Stalinism and dynastic monarchy, a tiny but potentially globally fatal residue of the Cold War. Nonetheless, on both sides, memories of past conflicts provide an irresistible ideological resource by which political elites, security agencies, and media demonize competitors in a system to which all subscribe. As Ilya Budraitskis (2014) remarks,

> the United States plays up its devotion to "liberty," appealing to Russian liberals whose Skype conversations with Western NGOs are recorded by the NSA, while Russia appeals to Western leftists (and Eastern Ukrainians) by capitalizing on nostalgia for the Soviet Union and the idea, more propagandistic than realistic, that state capitalism is markedly superior to the liberal variety. Too often, however, this is what defines the debate: each state's propaganda machine, with patriots believing their own country's talking points and dissidents believing the other's, obscuring what out to be the glaringly obvious fact that neither nation-state is motivated by any principle in domestic or global affairs more honorable than "what's good for our oligarchs," who even live in the same parts of Manhattan.

In this context, international disputes, as Houman Sadri and Nathan Burns (2010) observe, resemble not so much the Cold War as the Great Game of nineteenth-century intracapitalist diplomacy. It is these conflicts that have provided the second major axis around which cyberwar has unfurled—around issues from Chinese digital espionage, Russian cyberattacks on Estonia and Georgia, North Korean Hollywood hacks and bank cyber-heists, and U.S. malware implantation in its rival's networks.

The war on terror and the New Cold War were intensified and inter-twined by an event that Trotsky would have understood well—the Wall Street crash of 2008 and subsequent economic recession. The most serious economic crisis of capitalism since the 1930s raised class antagonisms to the level of what Franco "Bifo" Berardi (2016) calls "global fragmentary civil war." It ignited a sequence of heterogeneous popular uprisings run-ning from Tunis and Cairo to Madrid, London, New York, Istanbul, Rio de Janeiro, and Kyiv that variously mixed anticapitalist, liberal, libertarian, and ultra-right elements in revolt against oligarchy, corruption, unemploy-ment, and precarity. A notable feature of these movements was the use of social media for the mobilization of protest and occupation, earning the rebellions the dubious appellation of "Facebook revolutions," and also involving a hacker wing, exemplified by the WikiLeaks and Anonymous groupings that overtly declared cyberwar against established power.

The scale of the crisis, which came close to cracking the financial edifice of the global economy, and of the subsequent tumults, in which regimes fell to protest in streets and squares, had a dramatic effect on a political left in retreat since the end of the Cold War. It made it possible to think again in terms of large-scale systemic change. For some, it restored a "communist horizon" (Dean 2012). Even those who recoil from that phrase because of atrocities of Stalinism or Maoism could reconsider the prospect of a world beyond capitalism. But anticapitalist hopes were not fulfilled by the turmoil. Most protests failed, and in those that suc-ceeded, the ejection of governments was followed by the installation of new regimes as or more committed to neoliberalism. Occupations and assemblies movements did leave a legacy in new electoral experiments in socialist or social democratic politics: Podemos in Spain, the Corbyn Labour Party in Britain, and the Bernie Sanders campaign in the United States. However, in Europe and North America, failure of any broad left breakthrough opened the way for economic misery to be recouped by movements of the far right. Neofascists seized the moment for their own dramatic digital deployments and provocations, the most momentous instance being the role of the alt-right in the election of Trump in the United States.

All these processes intensified the drift and drive to war, and to cyberwar. In Ukraine, Syria, and Libya, with all the differences and similarities between these cases,[4] popular uprisings precipitated internecine civil and proxy wars and foreign military interventions, creating battlefield laboratories for digital as well as conventional weaponry. War in eastern Ukraine became the most dangerous focus of the so-called New Cold War between the United States and Russia, with savage fighting punctuated by hacked electricity-grid blackouts, virtual espionage, and torrents of computational propaganda. The Syrian civil war added a new catastrophe to the "war on terror," catalyzed the rise of ISIS, and also formed a new focus of confrontation between the United States and Russia and their regional allies. It also became the most network-mediated war history in terms of contending social media reportage, digital surveillance and deception, and hacking and counterhacking. The military defeat of ISIS in Syria and Iraq in 2017, in a conflict where drone warfare and cyberoperations were part of the repertoire of all contenders, seemed likely only to initiate a new round of conflict between regional powers. The tensions generated by war in Ukraine and Syria in turn provided motive for whatever hacking interventions Russia may have made in the 2016 U.S. election. The incoming Trump administration raised the long-brewing U.S. attempt to suppress North Korean nuclear weapon development, for several years fought out in terms of asymmetrical cyberstrategies, to the boiling point, in a crisis that implicated a China increasingly assertive and impatient with American global hegemony.

The increasing frequency and intensity of cyberwar reveal that a world dominated by the market is not necessarily peaceful or one from which the prospect of catastrophic conflict has been abolished. While neoliberalism not only acknowledges but enthusiastically celebrates competition and its associated "disruptions" and "creative destruction" as a spur to innovation and wealth creation, its official message is that, ultimately, the invisible hand of the market anneals this into the greater good of optimal resource allocation. What is denied in such discourse is the possibility that the conflicts of an agonistic and increasingly ahuman system might explode into noncreative destruction and that the ultimate "disruption" is to be found in the horror of war. Cyberwar is a secretive but increasingly irruptive manifestation of this possibility, a partially contained yet now escalating expression of the world market's destructive tendencies. Its rise may be a symptom of a new era of capitalist war, a pattern of covert but escalating conflict between great powers, and between these powers and the terrorist movements they have beckoned into existence, in a concatenation of conflicts running from Central Europe to the Middle East and the South China Sea.

It is this situation that Étienne Balibar (2015) described when, writing in 2015 from a European context, he observed,

> Yes, we are at war. Or rather, henceforth, we are all in war. We deal blows, and we take blows in turn. We are in mourning, suffering the consequences of these terrible events, in the sad knowledge that others will occur. Each person killed is irreplaceable.
>
> But which war are we talking about? It is not an easy war to define because it is formed of various types which have been pushed together over time and which today appear inextricable. Wars between states (even a pseudo state like "ISIS"). National and international civil wars. Wars of "civilization" (or something that sees itself as such). Wars of interest and of imperialist patronage. Wars of religions and sects (or justified as such). This is the great stasis or "split city" of the twenty first century, which we will one day compare to its distant parallels (if indeed we escape intact): the Peloponnesian War; the Thirty Years War; or, more recently, the "European civil war" that raged from 1914 to 1945.

And this prospect brings a terrifying ambiguity: that of a deep destabilization of the existing order, of the very type that has in the past created openings for new social protagonists and collective experiments, but also for disasters and atrocities whose potential scale today extends to nuclear species extinction. Any contemporary radical politics should be unsparing about the relation of revolution to war. The preface to the bourgeois revolutions of the seventeenth and eighteenth centuries was written in the ledgers of war debts that bankrupted absolutist monarchies and in the main chapters inscribed in blood spilled on battlefields from Naseby to Valmy. In socialist revolutions, the prelude to successful armed uprising has been a period of sustained capitalist self-destruction: the Paris Commune, the October Revolution, and the Chinese communist revolution all demonstrate this; if there are exceptions, they are relatively minor. One can say that the lesson of the twentieth century for revolutionary politics is that only capital can destroy capital: nothing other than its own massive apparatus of destruction is adequate to the task of utterly disrupting a dominant, gargantuan, consolidated mode of production.

We need to be very clear: there is in our analysis no issue of welcoming or hastening or urging the acceleration of this situation, only of facing conditions not of our own choosing. Being "interested" in cyberwar, interested in its "interest" in us, is, we think, an important part of that confrontation. In starting our discussion with a trenchant crypto-Trotskyism on that topic, we do not to signal any a priori affiliation to past models of vanguard politics; on that score, any movement toward a future beyond capitalism probably has as much to forget as to remember. But we are "interested" in cyberwar to understand both the defeat of previous forms of revolutionary politics and the possible emergence of new ones.

CRITICISM OF ARMS, ARMS OF CRITICISM

Our project requires an analysis in terms very different from those of official discussions of cyberwar, and its echoes in the media, discussions often cast in a language that assumes and imposes the inescapable necessity of a technocratic realpolitik and prepares and armors its audiences for trajectories of mounting crisis. Just as, in previous decades, we were habituated to apocalyptic prospects by an anodyne "nukespeak," so today we are being trained to a more creeping catastrophe by a militarized "cyberspeak" that promises us resilience against hostile intrusion through social media service upgrades, improved smartphones, and the steady surrender of individual and social freedoms to enhanced powers of the security forces. This numbing discourse must be broken open and defamiliarized.

The topic of cyberwar has, however, largely been ignored by critical social theorists, sometimes where one would least expect. The most ambitious and sophisticated of all recent scholarly attempts to conceptualize the scope and depth of the internet as a technosocial institution is in our opinion Benjamin Bratton's (2016) *The Stack: On Software and Sovereignty*. For Bratton, the concept of a "stack" (in computer science, a collection of data elements that must be accessed in a specific order) provides a metaphor for the architecture of the internet, envisaged as a layered series of terrestrial, platform, urban, communication, interface, and user operations. The combined interactions of these different digital levels, Bratton argues, now compose an "accidental megastructure" of global governance either destabilizing or reinforcing the spatial and temporal boundaries of the nation-state.

Yet what is striking is the limited attention this otherwise comprehensive, virtuoso examination of digital networks gives to that most sovereign of activities, war, and the possibility of what Bratton (2016, 298) glancingly refers to as "Stack versus Stack" conflicts, a topic that he defers as a matter for later study. This is all the more surprising because, as Bratton acknowledges in a footnote, the modeling of the internet as a "stack" of technosocial activities was pioneered within U.S. cyberwar agencies with very nonaccidental strategic agendas (441n8). One of the NSA documents released by Snowden, "Bad Guys Are Everywhere, Good Guys Are Somewhere" (Risen and Poitras 2013; Müller-Maguhn 2014), contains a diagram of the architecture of the internet in terms of an interdependent "stack" of geographical, physical network, logical, cyberpersona, and persona levels, derived from the work of national security intellectual Paul Rosenzweig (2012). The diagram is defined as a tool that enables the NSA to conceive how to scan and disrupt the operations of antagonists at multiple levels: "the stack" is a concept for cyberwar fighting. No account of networked sovereigns and subjects can now postpone reckoning with this reality, so we hope that our book can supplement Bratton's, even though it is written in a different political register.

There are important intellectual precursors of and contemporary resources for such a task. Paul Virilio ([1977] 1986) was addressing "cyberwar" decades ago (Wilson 1994). His "dromological" analysis of speed in advanced capitalism, with its emphasis on military operations, prefigures today's digital conflicts, as does his concept of "pure war" as a runaway and socially exhausting technoscientific militarization (Virilio and Lotringer [1983] 1997) and his later predictions of the socially disintegrative effects of an "information bomb" (Virilio 2002), though his somber totalizations—a "politics of the very worst" (Virilio and Petit 1996)—leaves little space for alternative logics and strategies of struggle. As important (and subject to much the same criticism) is Friedrich Kittler's war-centered theorization of media history, especially his account of the destructive genealogy of digital technologies.

There are also many current works that touch parts of the digital war elephant now trampling through cyberspace. Texts dealing with the emergence of digital capitalism (Edwards 1996; Schiller 1999, 2014) chart the original and ongoing role of the military–industrial complex in its creation. Powers and Jablonski's (2015, "Introduction: Geopolitics and the Internet") broader conceptualization of cyberwar as "the utilization of digital networks for geopolitical purposes in order to further a state's economic and military agendas" has been important to us. There is abundant work on the consolidation and mobilization of user populations across the corporate digital platforms that are a crucial matrix for cyberwar activities (Fuchs 2014; Chun 2016; Srnicek 2016). We have already mentioned the bourgeoning "big data" literature. Post-Snowden, there is also a fast-growing body of surveillance studies that examine this critical feature of cyberwar (Greenwald 2013; Schneier 2015). There is an even deeper range of literature on hacktivism—the dissident side of cyberwar—by both scholars (Jordan 2008, 2015; Hands 2011; Greenberg 2012; Olson 2012; Coleman 2015) and, of special importance, practitioners (Assange 2012; 2014). The pragmatically oriented work of digital civil liberties defenders like Ron Deibert (2013) and his collaborators on issues of censorship, surveillance, and network militarization is a vital resource. And there are now several critically important exposés of the war on terror (Mazzetti 2014; Scahill 2013; Turse 2012) and its connections to cyberwar issues via, for example drone warfare (Cockburn 2015; Shaw 2016) or the new military urbanism (Graham 2016). We learn from and draw on all these sources.

None of the aforementioned texts, however, singularly confronts the concept of cyberwar or addresses its consequences for contemporary subjectivity. Of texts that come closest to our perspective, Jodi Dean's (2010) *Blog Theory* is exemplary in its combination of Marxian and psychoanalytic perspectives to explore the compulsive power of "communicative capital-ism's" networks of enjoyment, production, and surveillance but hardly addresses the military dimensions of this equation. On the other hand, Brian Massumi's (2015) *Ontopower: War, Powers, and the State of Perception* speaks strongly to the present wartime conjuncture, in an impressive synthesis of political, philosophic, and neuropsychological thought, but arrives at conclusions different from our own. Éric Alliez and Maurizio Lazzarato's (2018) *Wars and Capital* and Andreas Bieler and Adam David Morton's (2018) *Global Capitalism, Global War, Global Crisis* anatomize the connection between war and the dominant mode of production but do not share our primary focus on the cybernetic aspect of this relation. As we finalize this book, further critically theorized works are doubtless in preparation, and a good thing too: *aux armes, citoyens!*

For our part, taking up Trotsky's purported injunction to grasp how and why "war," today, is "interested in you," this book develops a Marxist–Lacanian perspective on cyberwar, examining it both in political–economic terms, as a manifestation of the class antagonisms of global high-technology capitalism, and psychoanalytically, as a field where these contradictions are charged with fantasies and imaginary misrecognitions, production of affect and capitalization of fears, and symptoms of users. In our interpretation, the emergence of cyberwar is a manifestation of the "capitalist unconscious" (Tomšič 2015); that is to say, it is symptomatic of features of the world market unacknowledged in the neoliberal discourse of globalization, namely, the profound aggression and destructiveness intrinsic to an order predicated on privatized and increasingly machinic competition. This is the problematic we address in what follows.

We favor the term *cyberwar* over close synonyms such as *information war* or *netwar* because its emphasizes the new centrality to war of digital technologies, thus pointing back historically to origins in Second World War and Cold War cybernetics and forward to the new levels of network-ing and automation likely to characterize all social relations, including war making, in the twenty-first century. Precisely because of this common medium, we also think it is important to acknowledge the continuities between manifest military operations and realms of digital policing, security, and subversion, where, as Mark Neocleous (2014) argues, "war power" has never been cleanly and fully distinguishable from "police power"—or indeed, from the rebellious "counterpower" both strive to repress.

Thus, in contrast with the narrow, minimalist definition of cyberwar that focuses on the most overtly state-sponsored and manifestly materially damaging forms, we intentionally take a wide, maximalist optic. We see cyberwar encompassing a spectrum of operations from digitized propaganda through networked espionage to critical infrastructure attacks and including both battlefield digital systems and panoptic civilian surveillance. We consider it as waged not only by states but also, asymmetrically, by political insurgencies and social movements. And we think that to understand cyberwar, its overlaps and gray-zone intersections with cybercrime and hacktivism have to be grasped as a constitutive feature. We appreciate the argument that such an ecumenical definition may, by its very breadth, encourage state-led militarized "securitization" of digital networks (Hansen and Nissenbaum 2009) but think that, at a moment when such militarization is already well under way, it is more important to understand how cyberwar extends into and arises from not just Clausewitzian international conflicts but also structural violence, such as that exercised by capital in class war. In a time when cyberwar is rapidly becoming seen as the "natural" extension of a New Cold War between antagonized nation-state camps, we challenge the binaries and biases that this description reinstalls and the imaginary arrangements it produces to cover up the real connection between cyberwar and capitalism.

It is to understand the phantasmal energies that drive cyberwar and reckon with the way its "interest" in us virally implants at the heart of the everyday activities of the networked user the imaginary identities necessary for the prosecution of war that we invoke not only Marxist class and geopolitical analysis but also Lacanian psychoanalysis. Here we extend a line of Marxist–Lacanian media theory, already developed through the work Louis Althusser, Slavoj Žižek, Jodi Dean, and others, into the new digital wartimes. To combine Marx and Lacan is to work with both convergence and contradiction—convergence insofar as both decenter all concepts of an autonomous human subject, Marx situating that subject within the contradictions of capital's systemic exploitation of labor power, Lacan specifying its incessant, paradoxical division between conscious agency and the unconscious. In doing so, both insist on a deep, constitutive negativity, an otherness, at the core of the human condition. This negativity must be acknowledged and understood to overcome its destructive aspects of exploitation and psychopathology, aspects conjoined in the insanity of high-technology war. At the same time, we recognize profound differences in the political and psychoanalytic projects of Marx and Lacan but choose to find in them not just opposition but also a source of complementary strength. As Samo Tomšič (2015, 237) eloquently writes, "Lacan next to Marx questions the optimistic and humanist readings, according to which Marx's critique aims to break out of symbolic determinations, negativity and alienation. Marx next to Lacan questions the pessimistic and apolitical readings, according to which Lacan's . . . project supposedly amounts to the recognition of [a] 'universal madness' . . . that reveals the illusory nature of every attempt at radical politics." The "shared logical and political project" of Marxism and psychoanalysis is to "determine the terrain in which the subject is constituted" and "detach that subject" from the abstract, dominative, and fetishized forms capital imposes on it (Tomšič 2015, 237–38)—among which we count the forms of cyberwar's militarized networks.

In chapter 1, "The Geopolitical and Class Relations of Cyberwar," we develop a Marxist analysis of cyberwar, drawing not so much on the Leninist tradition of which Trotsky was a part as on more recent schools of open Marxism, autonomism, and communization theory. We examine how cyberwar is a manifestation of a larger metamorphosis of global capitalism, driven by its internal technological revolutions, and how underlying the international interstate conflicts with which cyberwar is usually associated are the deeper convulsions in which capitalism's populations are being compelled to labor in new ways but are also often becoming superfluous to its requirements. These complex, superimposed dynamics explode in the concatenation of interstate and interclass conflicts that followed the financial crash and economic crisis of 2008 and continue to intensify today.

In chapter 2, "Cyberwar's Subject," we look at how the ahuman logic of cyberwar operates at the level of the subject as a machinic capitalist unconscious in which the participants are made unwittingly, or at least unthinkingly, complicit by their everyday involvement in a ubiquitously networked context run by protocological and algorithmic processes. The logic of malware, virality, and mass metadata profiling yielding "hyperpersonalized" targeting (Dunlap 2014) makes it possible to be an unknowing agent and/or victim of cyberwar and creates new forms of mobilization and discipline—on one hand, a sort of networked *levée en masse,* and on the other, a martial panopticism targeting suspects and chilling networks. However, the increasing automation of these processes also makes humans peripheral and renders cyberwar subjects increasingly mystified and bewildered, a situation of confusion and misrecognition that paradoxically only intensifies the adoption of militarized practices and identities.

In chapter 3, "What Is to Be Done?," we consider the implications of cyberwar for current and future movements aiming beyond the society and subjectivity of capital. We distinguish four levels of problems cyberwar poses for such movements. The first involves questions of tactical practice under conditions of surveillance and hacktivism. The second entails new public issues of "fake news," criminal militarization, and escalating "dirty wars" (Scahill 2016). The third addresses problems of organizational form in a networked context. The fourth takes up the difficult strategic dilemmas confronting struggles that today are, of necessity, waged both in and against cyberwar.

① The Geopolitical and Class Relations of Cyberwar

IN THE LABYRINTH

Cyberwar is a murky domain with multiple and contested adjacencies to cybercrime, cyberespionage, and cyberactivism as well as to both conventional and nuclear war. Now we will make a way through this labyrinth, following a winding Marxist thread. For war is both everywhere in Marxism and nowhere. It is present in the frequent references of Marx and, especially, Engels to the conflicts of their age; in the writings of Lenin and Trotsky on war and revolution; in Gramsci's discussion of wars of position and maneuver as metaphors for political strategy; and in the development by Mao, Guevara, and Giap of the concept of people's war. Yet there is no definitive or canonical Marxist theory of war.[1] In an essay on "Marxism and War," Étienne Balibar (2002) confronts this issue, suggesting that while war has never been a *central* concept for Marxism of the same order as, say, "exploitation" or "commodification," it has always been a *problem* to which its thinkers repeatedly returned. The problem of war for Marxism is, Balibar says, that it requires the articulation of at least three different issues. The first is the class war between capital and labor; the second competitive wars between capitalist states; and the third the issue of revolutionary wars waged against capital. Balibar's schema suggests the need for a multidimensional understanding of cyberwar, connecting apparently disparate contemporary conflicts. However, in regard to "revolution," we suggest that cyberwar demands thinking two very different kinds of revolution: political revolution *against* capital and

33

technological revolution *within* capital. So we will work with four headings: class war, state war, revolutionary war, and technological revolution. As will rapidly become clear, each of these categories, though apparently discrete, actually contains all the others within it, so that by the conclusion of this chapter, we have an analysis of cyberwar in which these four approaches converge. But to arrive there, we will successively enter the cyberwar labyrinth through each of these four separate doors, starting where conventional accounts of cyberwar frequently begin: with technological revolution.

TECHNOLOGICAL REVOLUTION

That capital periodically renews itself with technological revolutions, machinic steroid injections that boost productivity, open markets, overcome tendencies to stagnation and declining profit rates, destroy old industries and create new ones, is an understanding shared by capitalist and communist theorists alike. Joseph Schumpeter's (1942, 139) rewrite of Marx and Engels's ([1848] 1964, 63) observations on the bourgeoisie's compulsion to "constantly revolutionize the means of production" as a paean to capital's "gales of creative destruction" exemplifies this concurrence. Capital comes into being as a socially radical force characterized by dramatic technological innovation and is then driven by competition between firms, and by those firms' need to defeat and disempower labor, to incessantly update its machinic apparatus, a process that periodically rises to the level of vast systemic convulsion, as in the first and second industrial revolutions of the early and late nineteenth century, based on steam and electricity, respectively.

In the 1970s, the accelerating adoption of computers and networks within advanced capital was recognized by Marxist theorists of the time as the "third industrial revolution" or the "microelectronics revolution" (Mandel 1975; Levidow and Young 1981). And just as previous industrial revolutions brought new iterations of industrial warfare, from dreadnoughts to tanks and aerial bombardment, so it is no surprise that the most recent of capital's technological revolutions manifests eventually not just in the means of production but also in the means of destruction, making cyberwar the logical military outgrowth of what is referred to as "information capitalism," "digital capitalism," "cognitive capitalism," or, indeed, "cybernetic capitalism" (Davis, Hirschl, and Stack 1997; Schiller 1999; Moulier-Boutang 2011; Robins and Webster 1988). However, to say just that cybernetic capitalism creates cyberwar—a commonplace—is too simple. For war comes first. In the majority of Marxist writing, capital's constant renewal of the means of production is seen as arising from normal market processes, that is to say, the rivalry between capitalists seeking competitive advantage, albeit advantage that may be won in class war by technologically breaking the power of the workers. But for a smaller group of theorists, the link between war and innovation is more direct.

Such a link is at least hinted at in Marx's (1973, 49) cryptic reminder to himself in the *Grundrisse* notebooks:

> Notabene in regard to points to be mentioned here and not to be forgotten:
>   (1) War developed earlier than peace; the way in which certain economic relations such as wage labour, machinery etc. develop earlier, owing to war and in the armies etc., than in the interior of bourgeois society. The relation of productive force and relations of exchange also especially vivid in the army.

This suggestion has been expanded by a number of heterodox Marxist historians who see armed force as crucially catalytic to capitalist development. Thus Robert Kurz ([1997] 2011) proposes that the seeds of capitalist modernity lie in the sixteenth-century "revolution in military affairs" arising from the discovery of firearms, demanding both large-scale military manufacturing and standing armies of paid soldiers—the "first waged workers." Peter Linebaugh and Marcus Rediker (2000) suggest that naval warships of the eighteenth century were not just maritime vehicles of mercantile expansion but provided a prototype for the disciplined organization of factory labor. Similarly, David Noble (1986), who carefully documented the links between industrial and military adoption of cybernetic technology, found precedent for this in the spurring of nineteenth-century machine production in the United States by the requirements of Civil

War–era arsenals and ordnance departments for large-scale, standardized arms production. All suggest that war is a root, not a branch, of capital's successive technological revolutions.

Following this line, we will say that the history of cyberwar is *not* just that war is transformed by cybernetic capitalism but rather that war creates cybernetic capitalism. A scrupulous reading of how "cyber" drives "war" will necessarily reveal that it is "war" that drives "cyber." The destroying force that propelled cybernetics was arguably more massive than any earlier example of the armed impulse to capitalist innovation, for it was provided by the mid-twentieth-century U.S. military–industrial complex forged in the world wars and the Cold War. *Cybernetics* has two meanings. The first designates a specific school of scientific thought that emerged during the 1930s and 1940s among researchers working on radar, ballistics, crypto-analysis, and atomic weapons for the U.S. and British war effort and that metaphorically adopts the Greek term *cyber*—for "governor"—to designate the concept of machines as entities governed by information control (Johnston 2008). The second, broader sense metonymically transfers the term to the entire realm of computer systems, from mainframes to mobiles, in whose evolution the work of the original cyberneticists played a crucial role. In both senses, cybernetics was born of war, whether as the fire-control systems for antiaircraft batteries developed by Norbert Weiner, the code-breaking computers developed by Alan Turing so central to intelligence operations, or the computing devices critical to calculating the effects of atomic fission pioneered by John von Neumann and others, vital to the development of nuclear weapons.

Thus war becomes "cyber" by 1945, well prior to other spheres of capital. This also means that at the root of cybernetics is not just geopolitical strife but class conflict. For if the development of computers by the United States and its allies was initially driven by the struggle against fascism, this propulsive institutional force was very rapidly taken over by the antagonism with the former wartime ally, the USSR, whose central role in the defeat of Nazism by 1945 positioned state socialism—however monstrously "bureaucratically deformed"—as a systemic challenger to the supremacy of U.S.-led capitalism. It is in the Cold War that the "cyber" really gets into "war," incubated within the U.S. "iron triangle" of military, corporate, and academic interests—which met the Pentagon's computing needs (Edwards 1996, 47).

Thus, as Noble and others have tracked, the computerization of industrial processes is initiated by military projects such as the Whirlwind computer, developed by MIT for the U.S. Navy as a flight simulator; the massive Semi-Automatic Ground Environment (SAGE) air-defense system, intended to protect North America from Soviet bombers; and the "secret empire" of signals intelligence and spy satellites of the NSA's "code warriors" (Taubman 2003; Budiansky 2016). Cybernetics also found its way onto "hot" battlefields of the Cold War, whose casualties are reckoned in millions. Operation Igloo White became a major part of the bombing of Vietnam, seeding the Ho Chi Minh trail with motion sensors communicating with a secret central control room in the jungles of Thailand where, behind airlock doors, IBM technicians processed a stream of dubious data on the supercomputers of the age to call in air strikes on troop convoys— or misidentified peasants (Edwards 1996; Cockburn 2015; Levine 2018).

Of particular importance to the computer industry was its deep connection to nuclear war preparation. Many consider MANIAC, the ironically named Mathematical Analyzer, Numeric Integration and Computer, developed to model the fission and fusion processes of the hydrogen bomb, the first modern computer with random access memory containing both data and instructions. The role of computers in simulating the extraordinarily complex dynamics of nuclear explosion continued to propel digital innovation and was of critical importance. As Blake Wood (2005) observes, from 1945 to 1975, when "all features of the modern nuclear weapons and many of the US stockpile devices were designed," the development of computer technology was "*driven* by the nuclear weapons program" (emphasis original).

Nuclear war also originates the internet. It is well known that Paul Baran developed the idea of digital packet switching working as a RAND Corporation employee on the problem of making military communication survivable in the midst of nuclear war (RAND 2008; Metz 2012). These principles were first actualized in the digital network developed by the Pentagon's Advanced Research Project Agency (ARPA), although they were used to connect computer facilities for scientists working on military-funded research rather than missile bases. Many computer scientists who worked on ARPANET have been keen to stress the autonomy of their research from directly military purposes (Hafner 1998). But as Janet Abbate (1999, 76) observes in her detailed account of the "invention of the internet," despite the perception of computer scientists and graduate students that ARPA doled out research funding with little concern for its application, military imperatives in fact drove the research agenda. Yasha Levine (2018) has recently documented how aware ARPANET's famous director J. R. Licklider, who had himself worked on U.S. nuclear war air-defense systems, was of the network's military priorities. ARPA allowed its employees considerable latitude, but as a sophisticated sponsor harvesting technological experimentation for techniques of annihilation.

Silicon Valley thus grew as a global center of digital industry supported by military contracts from firms such as Raytheon, IBM, and Sperry. It drew on a new type of scientific worker—the computer *hacker*, in the term's original meaning of a digital tinkerer or experimentalist—created in university departments funded by military research. Many accounts of the information revolution tell a "swords into plowshares" story in which the internet is rapidly hacked free from military tutelage by system administrators and students at large universities, liberated for use by researchers and countercultural experimenters, and then by commercial developers, who take the digital on a fast march away from governmental supervision. This narrative is partially true. From the 1980s on, the digital economy *was* increasingly privatized, deregulated, and directed toward meeting business needs and creating consumer commodities (Schiller 1999). Today corporations like Google, Amazon, Apple, and Facebook exercise a level of control over technological innovation and policy that seems to surpass that of the government that initially seeded the computer revolution. But focusing on the escape of digital capital from its martial incubator obscures the continuing involvement of the "iron triangle" (Edwards 1996, 47) of military, industrial, and academic interests in internet and computer development, both as an instigator of research from which civilian innovation is spun off and, in turn, as recipient of civilian technologies adapted to military purposes. And this is so through the entire progression of "cyber," from "first-order" cybernetics as a system of automation (such as the antiaircraft fire systems Weiner worked on) to "second-order" cybernetics as networks (ARPANET) and "third-order" cybernetics as automated algorithmic networks (such as the surveillance systems of today's NSA) (Bousquet 2009; Rid 2016).

Indeed, though the nuclear weapon systems to which computers and networks were so integral were not used in war after 1945, it was arguably a form of "cyberwar" that eventually won the Cold War for capitalism. New generations of first-strike nuclear weaponry developed by the Reagan administration—cruise missiles, Trident submarine-based intercontinental ballistic missiles, and the Strategic Defense Initiative "Star Wars" systems, all deeply dependent on computerized guidance, navigation, and targeting (Scheer and Zacchino 1983; Aldridge 1999), drew the Soviet Union into what it found an unaffordable arms race. Its planners confronted an intractable choice between "guns or butter": more high-tech weaponry or the consumer goods demanded by an increasingly dissatisfied populace. This direct military pressure was in turn part of a broader economic defeat, as the USSR's industrial strengths were overtaken by new informational forces of production spun off from the Pentagon matrix and rapidly germinating in Silicon Valley (Shane 1995).[2] The USSR's cyberneticists, fatally hampered by a repressive state apparatus, failed to match developments not only in personal computing but, even more strikingly, in networks (Spufford 2010; Peters 2016). It is on the field created by the defeat of the state socialist project that there would emerge the new interstate hostilities driving today's cyberwar.

INTERCAPITALIST CYBERWAR

Marxism's concept of imperialism asserts that at any given historical moment, the world market is led and dominated by a specific nation-state that concentrates within itself the most advanced powers of capital, until it is challenged and defeated by emerging rivals—typically in a period of cataclysmic wars.[3] This idea, central to the thought of Lenin (1939) and Mao ([1938] 1967), also informed academic "world-system" theory, mapping a planet dominated by a "first world" core of imperial advanced capitalism, led by the United States, surrounded by states of the "second" and "third" worlds, condemned to exploitative underdevelopment unless freed by revolution (Amin 2010). That cartography was in one way confirmed by the United States's Cold War victory. In others, however, it was deeply disrupted, both by the defeat of state socialism and by the third world's fissuring between failed states and marketized success stories, preeminently that of China, now predicted to overtake the United States as the world's largest nation-state economy within decades. In grappling with a recapitalized, "postsocialist" world, Marxism has, moreover, divided between schools emphasizing the unity of the capitalist system as a whole and those stressing its competitive antagonisms. Michael Hardt and Antonio Negri's (2000) account of a planet where nation-based imperialism is superseded by the totality of a multinational capitalist "Empire" exemplified the former perspective. However, other theorists see signs of the renewal of cyclical struggles for imperial domination (Arrighi 2010) or fluid contests for hegemony waged by rival state-led blocs of capital (Worth 2005, 2015). Today's cyberwars demonstrate this latter, conflictual trajectory.

The triumph of U.S.-led liberal capitalism over state socialism was won both by the superiority of computerized nuclear weapons systems and by the attractions of a consumer capitalism revitalized with information technologies. Yet the apparent smoothness of the "digital Pax Americana" (Segal 2016, 223) was shot through with contradictions, for the very globalizing drive of cybercapital placed the elementary weapons of cyberwar—computers and networks—in the hands of a multiplicity of potentially rival state actors, all situated within a capitalist world market but each advancing the interests of particular corporate entities and specific fractions and blocs of capital. There are variously estimated to be between four dozen (Klimburg 2017) and 140 states (Suciu 2014) capable of waging some form of cyberwar. Many are allies of the United States, but others are opposed to its hegemony.

It is sometimes suggested that cyberwar levels the playing field between large and small states, and between states and nonstate actors, by lowering the cost of havoc-creating technology to that of a computer and a network connection. There have certainly been notable exploits by minor agents against major cyberpowers.[4] Social movements and insurgencies, from alter-*mondialistes* to al-Qaeda, from Anonymous to ISIS, now regularly challenge state power online. Proliferating virtual weaponry contributes to the chaotic volatility of real or suspected cyberattacks. But the idea that cyberwar equalizes international power is mistaken. As Alexander Klimburg (2017, 304) makes clear, high-level cyberwar operations today not only demand large teams of operatives with specialized skill sets but also depend on an extensive "logistics base" including malware arsenals and "cyberfoundries," where new weapons are developed and tested; "targeting aids," ranging from intelligence about opponents' networks to "implants already pre-positioned in the target system"; and "transport and forward operating bases," such as "rented botnets" or "totally subverted larger private networks" used as staging grounds for attack. In cyberwar, God continues to favor the big battalions.

Assessments of state cyberwar capacities therefore often rank nations in a hierarchy, depending on their abilities to combine defense, attack, and espionage (Klimburg 2017). The spectrum runs from states in Southeast Asia or Latin America with rudimentary defenses and dependence on criminal or mercenary proxies for offensive capacity to those developing national powers of defense and attack, slowly (Pakistan, India, Brazil) or rapidly (North Korea, Iran, EU states) through those that have already consolidated such powers (France, South Korea, the Netherlands, Switzerland) to others with "extensive offensive capacities," such as the United Kingdom, Israel, Russia, China, and, at the apex of the hierarchy, the United States. Elsewhere we address events involving Israel, Iran, Saudi Arabia, North Korea, and other mid-level cyberwar states. Here we offer only a snapshot of the capacities of the United States, China, and Russia, the three states whose cyberwar powers are most advanced and whose escalating hostilities are most widely feared.

The United States, an imperial hegemon that, however recently dented, remains militarily supreme, is in "a class of its own" by virtue of the domestic technological prowess and vast budgets it brings to cyberwar (Klimburg 2017, 305). After Cold War victory, the Pentagon's original coupling of computers with war morphed into a "Revolution in Military Affairs" that assigned a central battlefield role to smart weapons, such as precision guided bombs and missiles, and the intelligence-gathering and communication-disrupting power of information technologies.[5] The efficacy of this doctrine was demonstrated, at least to the satisfaction of defense budgeteers, in wars in Kosovo and Serbia in the 1990s, and then by the (seemingly) speedy invasions of Afghanistan (2001) and Iraq (2003). However, while the new digital battlefield systems were widely publicized, a more covert process saw the development of truly networked weapons. When, in 2009, the U.S. Department of Defense, after years of military–bureaucratic infighting, created "USCyberCommand" and declared "cyberspace" as a military domain equivalent to land, sea, air, and outer space, this was presented as a response to mounting hostile incursions. However, while the new entity certainly did aim to strengthen U.S. cybersecurity, behind the narrative of homeland defense lay programs for offensive cyberweapons.

This was made clear with the discovery of the Stuxnet "worm" that, between 2010 and 2012, disabled over a thousand centrifuges at Iran's uranium enrichment plant outside Natanz to prevent the building of nuclear weapons. Iranian scientists for a long time misrecognized the cause of the malfunctions because of Stuxnet's impeccable simulation of mechanical failure apparently unrelated to software performance. Stuxnet's software was complex, exploiting difficult-to-find, or expensive-to-buy, vulnerabilities. It is estimated to have taken eight to ten people six months to write and required laboratory testing and extensive intelligence gathering to effectively target (Schneier 2013). It is generally held to be have been designed and created in the mid-2000s as a joint project of U.S. and Israeli intelligence services. As the first confirmed example of an infrastructure-disabling cyberattack, unleashing Stuxnet was a "Rubicon-crossing" (Zetter 2014a) moment in the annals of digital warfare.

Reports of other U.S. cyberoperations planned against Iran and North Korea, and of the threats made by the Obama administration in response to Russian 2016 election interference, confirm a picture of far-reaching U.S. cyberwar-fighting posture (Schneier 2013; Entous, Nakashima, and Miller 2016; Arkin, Dilanian, and McFadden 2016). Snowden's revelations not only disclosed the extensive surveillance programs that are an integral, "defensive" part of U.S. cyberwar capacities but also a buildup of offensive powers by elite hacker units, such as the NSA's Tailored Access Operations (TAO) TURBINE project for the automated malware infection of millions of computers around the world with implants for surveillance and sabotage (Gallagher and Greenwald 2014).[6] Another Snowden-released document records that in 2011 alone, the United States launched 231 operations to infiltrate and disrupt foreign networks (Gellman and Nakashima 2013).[7]

The development of these capacities cannot be seen in isolation from other aspects of U.S. war planning. In 2014, the Pentagon announced an "Offset Strategy" to regain a military superiority it claimed was being eroded by rival powers by "harnessing a range of technologies including robotics, autonomous systems and big data . . . faster and more effectively than potential adversaries" (*Economist* 2018). The historian Alfred McCoy (2017, "Information and the Future of American Domination") situates offensive cyberwar weaponry as one part of a "triple canopy" strategy of U.S. power projection, the other components of which are armed drones and ubiquitous advanced surveillance. U.S. cyberwar operations are now deeply integrated with, and indeed inseparable from, conventional, kinetic U.S. military operations in Afghanistan, Iraq, Syria, and elsewhere (Gjelten 2013; Harris 2014). All this makes cyberwar a major element in U.S. imperial ascendancy.

It is sometimes suggested that the U.S. military's prowess in digitally attacking "hard" targets, be they tanks, bases, or industrial infrastructures, leads it to overlook the psychological and propagandist aspects of cyberwar (Klimburg 2017, 134–58). This view neglects the considerable resources of "soft power" wielded not so much by the Pentagon as by the State Department. These resources include close relationships with digital corporate giants, such as Google and Facebook (Fattor 2014; Powers and Jablonski 2015). The doctrine of unimpeded free access to the world market is not just that of the giant U.S. media companies but also that of the U.S. state, precisely because of its symbiotic relationship with its own large-scale capitalists, who in turn have an agenda for the deepening privatization of governmental functions. Signing on to Facebook or Google gives users access to planetary digital culture via a tacit "clicking" acceptance of Silicon Valley corporate entrepreneurialism, globalizing marketization, and an ethos of deregulated technological accelerationism, accompanied by the vast gatherings of personal, social, and territorial data (Lovink 2017).

These companies' collaborations with the U.S. security apparatus include their acquiescence to the NSA PRISM surveillance program, an episode they are now energetically attempting to live down. It also includes their lucrative provision of services to U.S. military and intelligence agencies, ranging from massive data storage and processing facilities (Amazon and Google) to space-based "geospatial visualization services" (Google) to "wearable tech" for the armed services (Apple) as well as surveillance systems for the private military contractors that now constitute the sizable penumbra of Pentagon operations (Alexander 2015; Gregg 2017; Levine 2014a, 2014b).[8] This reciprocity also involves the "revolving door" between high-level Silicon Valley and Pentagon managers and their shared culture of global "problem-solving" think tanks, typified by Google's Jigsaw, and a willingness to censor internet material at the behest of the U.S. government, once only sporadically disclosed but now in full swing after the 2016 election hacking scandal (Assange 2014; Levine 2014c; Greenwald 2017b). Yet the United States has cyberwar vulnerabilities. Paradoxically, these result from its historically advantageous "home game" (Segal 2016; 35) advantage: a highly networked military and economy susceptible to hacker attack. This problem is intensified by the neoliberal form of capitalism the United States champions. Free market zealotry is at once governmentally embraced and on occasion makes it difficult for the U.S. state to enforce cybersecurity provisions on private capital, around issues ranging from the reporting of cyberattacks to mobile phone encryption. It is largely in reaction to U.S. strengths and weaknesses that Russia and China have developed their own capacities and strategies. In particular, rather than seeing cyberwar as a distinct sphere of military operations, it is for these nations' military forces subsumed within a larger category of "information warfare" that comprehends forms of psychological operations and propaganda. This orientation, taken together with the different historic levels and paths of technological development, has led to distinctive approaches to conflict in cyberspace.

The People's Republic of China (PRC) was until recently the main object of popular U.S. cyberwar anxieties: depictions of transpacific wars fought with digital weapons abound in both fictional entertainments and serious political scenarios (Singer and Cole 2015; McCoy 2017). Yet, although China's digital policies and practices have been presented in the most ominous terms in U.S. media, they may be as much a sign of weakness as of strength (Lindsay 2014; Lindsay, Cheung, and Reveron 2015). An important factor driving the leaders of the Chinese Communist Party to open their country to the world market at the end of the 1970s was the perception of massive and intensifying technological backwardness vis-à-vis the West. As the PRC confronts the question of whether, in the global market, it will be an economic "head servant" (Hung 2009) to the United States or risk conflict over issues of trade and political dominance in the Pacific, this awareness shapes China's concept of cyberwar.

Thus, for the Chinese state, a major concern is, notoriously, maintenance of ideological control over a population perceived as vulnerable to terrorism, to separatism, and to "foreign influences" of the sort projected by U.S. digital "soft power." This control is exercised through the elaborate and continually evolving apparatus of digital censorship, surveillance, and preemption shorthanded as the "Great Firewall of China." This system combines blackouts of proscribed websites, algorithmic filtering and monitoring of email, and interventions by internet militias (the so-called Fifty Cent Army, named for the alleged pay rate for each posting) or bureaucrats to divert or disrupt controversial discussions with pro-government cheerleading with digital surveillance of, and attacks on, overseas sites and groups attempting to evade censorship (Lindsay, Cheung, and Reveron 2015; Klimburg 2017). Administered via the collaboration of private-sector internet service providers with state authorities, the system has been characterized by a changing, erratic enforcement that by its very uncertainty increases the risk to dissenters. The Firewall, which has undoubtedly contributed to China's unenviable record of large-scale imprisonment of social activists and journalists, has been a frequent target of Western criticisms for violations of internet freedom, though these have subsided somewhat post-Snowden. China's possible larger plans for an extensive "social credit" system based on monitoring of internet practices and "scoring" of social and perhaps political behavior would represent the logical consolidation of this system (*Economist* 2016a).

The overlap between issues of hard and soft power in a cyberwar context is demonstrated by what Bratton (2016, 112) terms the "Sino-Google war." Between 2002 and 2010, Google was involved in protracted and complex disputes with the PRC over that state's internet censorship. These intensified when, in 2006, Google set up a Google China subsidiary, headquartered in Beijing. After oscillating between compliance and noncompliance, and having suffered highly sophisticated hacking attacks, which the NSA offered to help counter (Levine 2014b), Google China closed its mainland office in 2010 and, though it continued operating from Hong Kong, now holds only a tiny share of China's search activity. This episode can be interpreted in several ways: as a clash between internet freedom and authoritarianism (the version favored by Western politicians, internet pundits, and media); as an intercapitalist dispute between Google and China's major domestic, and state-championed, search enterprise, Baidu; or, as Bratton describes it, as a collision between rival concepts of the internet's relation to the state, one giving the former primacy over the latter, the other reversing that priority.

Although there is something to be said for all these perspectives, they overlook the one Bratton explicitly rejects: "superpower" conflict. Yet the period of the "Sino-Google war" was one of mounting tensions between the United States and China, including the unfolding aftermath of China's earlier capture of a U.S. spy plane (Hersh 2010); American accusations of large-scale digital espionage, both economic and military, by Chinese hackers (Segal 2016); NSA "implants" in the networks of China's giant telecommunications company (Hsu 2014); and widespread expectations that Sino-American tensions could trigger the first full-scale cyberwar (expectations still widely held, even if Russia has lately publicly supplanted China as an object of U.S. suspicions). In this context, it is hardly surprising that the issue of a giant U.S. internet company not only skirting China's Great Firewall but also amassing massive information about China's data subjects, and establishing itself as gateway to global networks, was extraordinarily tense.[9]

The more externally aggressive face of China's cyberwarfare has been espionage against both state and corporate targets in the United States and other Western nations. Chinese hackers have been held responsible for some very large data breaches both in the U.S. Department of Defense and among military contractors. It is difficult to tell how much of this might be caused by state operatives, including People's Liberation Army units, free enterprise cybercriminals, or patriotic hackers; indeed, the Chinese state itself may struggle to contain the proliferation of cross-cutting hacking initiatives. To the degree that such espionage is state sponsored, it seems to be aimed at information to bolster military and technological progress: as one commentator remarks, China has few such secrets the United States needs to hack (Lindsay 2014). How far the thefts translate into real industrial successes is dubious. Such hacking has abated since 2016, when the Obama administration complained vehemently, suggesting it may not have been rewarding enough to warrant major confrontation.[10]

The People's Liberation Army has, however, contemplated information warfare against a technologically superior adversary (i.e., the United States) for many years (Pufeng 1995; Liang and Xiangsui 1999). It operates special cyberwar units and espionage for specifically military purposes (Segal 2017). It is likely that some senior officers see disruptions of an enemy's digital networks as a decisive factor in future armed conflict, an "assassin's mace" to disable an otherwise more powerful antagonist by surprise offensive (Hambling 2009). Some observers believe China's abilities to mobilize large numbers of militia-style hackers for auxiliary military purposes (Green 2016) to be a serious threat to its opponents. Others, skeptical about such forces' technical proficiency, warn the greater danger may lie in a misplaced belief by China's military in a first-strike digital strategy unlikely to succeed but certain to bring retaliation (Lindsay 2014).

Although China has loomed large in U.S. cyberwar preparations, the Russian Federation recently displaced it as military bête noire; the exercise of its cyberwar capacities has certainly been more dramatic. Despite the USSR's overall failure to match U.S. networked computing capacities, it had strong electronic battlefield training, outstanding espionage and special operations units, and an important body of military doctrine on the importance of propaganda and psychological warfare in conflict with imperialist powers (Thomas 2000). Following the collapse of the Soviet regime and "shock therapy" marketization, Russia's armed forces were demoralized and debilitated. However, with Putin's reassertion of state powers and nationalist ideology, this legacy of information warfare was reactivated, encouraged by the security-force background of the president and many of his regime's "new nobility," as Putin's successor on the position of director of the Federal Security Service (FSB) described the members of the principal security agency of Russia (Soldatov and Borogan 2011, 4–5). This took place in a context of an intensifying friction with the United States and NATO over military and economic influence, most notably in the former Soviet republics but also more widely.

In 2013, a paper by the chief of Russia's General Staff, General Valery Gerasimov (2013), gave consolidated expression to Russian military anxieties about U.S. expansion in the preceding decade. It depicts this as proceeding not only through regime-changing interventions in Yugoslavia in 1991, Afghanistan in 2001, and Iraq in 2003 but also through "color revolutions" in which pro-Russian governments were ejected by nationalist, but also broadly pro-Western, liberal uprisings. These revolutions—in Eastern Europe, Central Asia, and, during the Arab Spring, the Middle East—are in the eyes of Gerasimov and other senior Russian military officials fostered by the United States and NATO through media such as the BBC and CNN; NGOs and social movements assisted by the U.S. State Department, the U.S. Agency for International Development (USAID), and the National Endowment for Democracy; and, most recently, social media in "Facebook Revolutions" (Cordesman 2014; Bartles 2016). While Gerasimov's proposed answer includes a buildup by Russia of high-technology weaponry similar to that of the United States, it places special emphasis on combining conventional forces with special operations, diplomatic and economic pressures, and activities in "information space," with the latter elements predominating over the purely military ones in a ratio of four to one, a mix that has been widely dubbed "hybrid" or "nonlinear" warfare (Galeotti 2014).

The so-called Gerasimov doctrine articulates a gradual evolution of Russian military practice, through a series of events that in Western eyes are often seen as flaming beacons of approaching cyberwar: the 2007 cyberattack on Estonia, precipitated by a conflict over the removal of the Bronze Soldier monument; the Soviet World War II memorial in Tallinn; digital attacks combined with military operations in the 2008 Russian intervention in Georgia (Deibert and Rohozinski 2012); and (shortly after the publication of Gerasimov's paper) a stream of hacks and digital propaganda operations in the long crisis of Ukraine, from the 2013 Maidan uprising through the 2014 annexation of Crimea and the protracted war in the Donbas (Darczewska 2014a, 2014b; Greenberg 2017b). Although in the West, this sequence is often interpreted as a manifestation of carefully thought out Machiavellian strategy, its development may have been more haphazard and experimental (Patrikarakos 2017);[11] however, a characteristic repertoire of tactics, variously permutated, is recognizable.

These include the digital dissemination of pro-Russian information and disinformation, closely orchestrated with that on other media outlets, such as Russia Today and Sputnik News, by "troll armies" either by paid (albeit at arm's length) state employees, most notoriously from the St. Petersburg Internet Research Agency, or by patriotic hackers for whom the Russian state could disavow any direct responsibility (Morozov 2008). These human agents are often combined with chatbots for automated propagation of messages. Alongside these operations, harder forms of hacking occur, including defacements of government and corporate websites, intrusions into and information thefts from networks of governmental and private-sector institutions, or disabling malware attacks on media systems or electricity grids. All of this can occur alongside conventional, paramilitary, or proxy military actions, which digital propaganda denies, justifies, or mystifies with an indecipherable multiplication of conflicting stories. These external manifestations of information or cyberwar are set alongside the Russian Federation's authoritarian regime of domestic governmental information control that, while not as comprehensive as China's, combines extensive targeted internet surveillance of dissenters and investigative journalists (at high risk from state security forces and shadow assassins) with blacklisting of selected websites for "extreme content" and the self-censorship and regime promotion of private media outlets owned by state-aligned oligarchs (Soldatov and Borogan 2011; Maréchal 2017).

The relations between the U.S., Chinese, and Russian states are not symmetrical: for all the West's loud alarms about Russia's trolls or China's hackers, it is unlikely that those states' resources currently match the long-accumulated and deeply budgeted technoscientific cyberwar expertise of the United States, so strikingly demonstrated with Stuxnet. Moreover, the clashes of the United States, China, and Russia by no means cover the rapidly expanding "dark territory" of contemporary interstate cyberwar (Kaplan 2016). Their conflicts intersect with those of their respective allies: Israel, Saudi Arabia, Iran, Palestine, North Korea, Pakistan, India, and other nations engaged in invisible battles with autonomous and independently dangerous dynamics. Nonetheless, our survey of the three great cyberwar powers gives some sense of the arsenals of virtual weaponry states currently deploy, the interdependence of cyberoffense and cyberdefense, and the intensifying fusion of cyber- and conventional warfare.

It also charts a triangle of state relations from which many today fear there is emerging the long-unthinkable possibility of armed conflict between great powers, "the next war" (*Economist* 2018), to which today's cyberwars would be a prelude. As we have already indicated, characterization of the current situation as a "New Cold War," with its invocation of an ideological collision of rival world projects, is misleading. The contenders are capitalist regimes, or regimes fast becoming capitalist, including both the victors of the old Cold War and the postsocialist states in which the communist project, widely corrupted by ruling elites who turned the state apparatus against their own people, capitulated to or compromised with the world market. They are encounters between a still dominant imperial hegemon (the United States) and rising (China) or declining (Russia) rivals.[12] If cyberwar is cold war, it is intercapitalist cold war, a semicovert manifestation of hostilities fought out, no longer between imperial capital and state socialism, but between differently nuanced but commonly oligarchic blocs of neoliberal (the United States), kleptocratic (Russia), and authoritarian state (China) capital, in which the cybernetic weapons that once gave the United States its dominance are turned against it by competitors, ascending or descending, in conflicts that spectrally reanimate the sentiments of the capitalist/socialist hostility, even while the protagonists are commonly subsumed within a system of global commodification. It is a harvest of dragon's teeth.

CLASS CYBERWAR

The world market is, however, not just a site of state conflict. It is also a vast field of frictions, sometimes explosive, sometimes silent, between capital and labor—an arena of class war. To miss this aspect of cyberwar is to fall into a conventional view of international politics as a chess game between competing national powers (Bonefeld 2006). Behind and within the contests of contending states lies a deeper set of conflicts; subtending and shaping the geopolitics of cyberwar is its role in the war of capital and labor, and in this war, too, virtual weaponry is wielded by both sides.

The transfer of computers and networks from their military incubators to the civilian workplaces of North America and Europe was spurred by economic crisis. By the 1970s, the strike power of Fordist industrial workers was driving wage and welfare gains even as intercapitalist international competition was intensifying. With a relentless logic, the Pentagon's new technologies, developed to fight state socialism, were switched to the home front. Cybernetic class war, waged from above, automated many manufacturing and office jobs; sent others offshore via telecommunication-controlled supply chains; and redirected profits to a financialization dependent on electronic stock markets, computer risk modeling, and high-speed algorithmic trading (Schiller 1999; Dyer-Witheford 2015). Over some forty years, capital's "cybernetic offensive" (*Tiqqun* 2001) broke the factory bases of the relatively well-waged mass worker of the planetary Northwest.

There have been many Marxist attempts to describe the new, post-Fordist class composition that emerged. Michael Hardt and Antonio Negri (2000) suggest the mass worker has been replaced by "immaterial labor" involved in digital networks. We are critical of this formulation; it overlooks both the shift of industrial labor to Asia (where it drove China's rise as a great power) and the generation of "surplus populations" thrown out of work in Rust Belt cities or inhabiting regions largely bypassed by digital supply chains, such as large sectors of Africa and the Middle East (Dyer-Witheford 2015). But in regard to cyberwar, the "immaterial labor" thesis is important, because it highlights the centrality to digital conflict of a new type of technoscientific labor that saw itself not as "worker" but as "hacker."[13]

Military production was the birthplace of the hacker. The famous "hacker ethic" of innovation, openness, empowerment, and belief that "information wants to be free," with its libertarian scorn of bureaucratic regimentation and corporate "suits," was the ethic of experimental systems administrators and adventurous graduate students working on U.S. Defense Department university contracts (Levy 1984; Himanen 2002; Wark 2004). This young and overwhelmingly male hacker workforce (and its legends) flowed out into a still largely Pentagon-bankrolled Silicon Valley and thence into a wider digital economy. As it did so, hacking split into different lines. The dominant line was corporate, professional, and entrepreneurial. It led, via Bill Gates's assertion of intellectual property rights in software, not only to Microsoft but onward to the corporate empires of Apple, Google, and Facebook. A minority trajectory pursued free software, cypherpunk encryption, digital commons, and the noncommercial distribution of network protocols. A third, subterranean break-off took hacking to profitable crime, raiding for credit card numbers, bank access, industrial secrets, and saleable software, a project that would attain a global scale. These threads constantly entangled with one another, frustrating attempts to find in the hacker a consistent politics, be it progressive or reactionary.[14]

All have a place in the history of cyberwar. Commercialized and professionalized hacker labor and entrepreneurialism drove software and network companies fueled by military contracts. Criminal hacking, though pursued and prosecuted by national security states' policing arms, also supplies these states' cyberwar with black market weapons, such as previously unrecognized software vulnerabilities known as "zero-day exploits" or dual-purpose criminal–military botnets. It also entered into an ambivalent revolving-door relation with cybersecurity firms, fluidly swapping black and white hacker hats. And from the minor line of free software and cypherpunk activism, and its meeting with antiauthoritarianism politics, came the connection between hacking and oppositional social movement: "hacktivism" (Greenberg 2012).

The first of successive "firebrand waves of digital activism" (Karatzogianni 2015) sprang up in the 1990s within an alter-*mondialisme* protesting the negative consequences of neoliberal globalization. Primarily a North American and European movement, but with major connections to India and Latin America, counterglobalization brought the immaterial labor of antiauthoritarian hackers into contact with the very material concerns of industrial workers losing their jobs and peasants losing their land. One of its starting points was the use of computer networks by insurgent Mayan Zapatistas to publicize their armed resistance to free trade agreements between Mexico and the United States. The famous announcements by RAND consultants John Arquilla and David Ronfeldt (1993, 1996) that "cyber war is coming," an early warning to the U.S. state that computer networks could be a medium of popular mobilization, were inspired by the eruption of "Zapatistas in cyberspace." Digital circulation of Subcommandate Marcos's poetic calls for resistance to neoliberal policies worldwide galvanized a "cyberleft" (Wolfson 2014) enabled by the increasing availability of personal computers and internet connections and the online experiences of young people versed in video games, music piracy, and the World Wide Web. Summit-busting demonstrations, from Seattle to Genoa, were accompanied by indie media centers, the digital relay of information among activists, and distributed denial-of-service (DDoS) attacks on corporate and state websites. Julian Assange honed his hacking skills as an "alter-globalist." Electronic civil disobedience, digital whistleblowing, and virtual organizing wove what Harry Cleaver (1995) termed an "electronic fabric of struggle."

Then the tide of alter-globalization suddenly ebbed. The main cause was the chilling effect of the 2001 attacks on the World Trade Center and the subsequent "war on terror." The decline of the cyberleft also, however, coincided with the U.S. dot-com crash of 2000, in which attempts at corporate appropriation of the net expired in a sea of red ink and stock market scams, as innumerable sketchy start-ups failed to find a business model to capture networkers used to free content. This crash might have strengthened the anticapitalist movement. It was, however, contained by the U.S. Federal Reserve Bank's drastic lowering of interest rates (a measure that would later boomerang in the much larger housing crash of 2008). In the dual meltdown of "dot-coms" and "dot-communists," the former recovered first. The crisis of U.S. digital capital winnowed winners and losers from the excess of a speculative boom, refining the strategies of fresh entrants to the field and inaugurating a new phase of internet history.

After a short hiatus, cybernetic capital rebuilt, with a new business model, "Web 2.0" (O'Reilly 2005). The technologies that enabled this transformation had strikingly varied origins, coming from both sides of capital's class cyberwar. Twitter has its origins in TXTMob, an application first developed by the Institute for Applied Autonomy for the self-organized coordination of protestors at the 2004 Democratic National Convention in Boston and the Republican National Convention in New York City (Radio Netherlands 2013). On the other hand, Google Maps grew out of the acquisition of Keyhole, a small Silicon Valley company supported by venture capital from the CIA's venture capital front company In-Q-Tel that worked to "develop fast, accurate and searchable digital maps for the US Armed Forces" (Powers and Jablonski 2015, 84). As Mariana Mazzucato (2013) has shown, the research behind almost every component of Apple's iPods, iPhones, and iPads was funded almost exclusively by government agencies, predominantly by the U.S. Department of Defense: in 2014, "the parent company of Siri's creator, which was acquired by Apple in 2010, still [got] over half of its revenue from the Department of Defense" (Bienaimé 2014).

The outcome of capital's omnivorous appetite for innovation was a dramatic revision of digital political economy and usage. The key was recuperation of internet aspects that had frustrated the dot-coms and energized the cyberleft: popular preference for conversations over published content and free over paid content. In Web 2.0, these seemingly subversive elements were mobilized for accumulation. The digital enterprise was reconceptualized as not "publisher" but "platform," managing proprietorial software that offered users a launch point and tools for structured but self-directed network activities (Bratton 2016). Monitoring and measurement of these activities supplied data for the algorithmic targeting of advertisements. Google, Facebook, and Twitter were flagships, but other businesses adopted elements of the model: Apple made its hardware a platform for apps and music; Amazon algorithmically recommended an ever-mounting heap of retail products. As oppositional energies declined, and "platform capitalism" (Srnicek 2016) burgeoned, driven by the free labor of user-provided content and the big-data flows of surveilled self-revelation, leftist digital optimism was replaced by Jody Dean's (2009) diagnosis of a "communicative capitalism" fully capable of commodifying the compulsive loops of so-called social media.

It was therefore startling when the economic crisis of 2008 brought a return of class struggle cyberwar. Wall Street's subprime mortgage crisis, relayed around the world by some of the most advanced computer networks in existence, had brought the global economy to a brink from which it was only hauled back by the massive state intervention of bank bailouts and austerity budgets. Responses from below differed in specific zones of the world market. Nevertheless, by 2011, Eurozone anti-austerity revolts, strike waves in China, the Arab Spring, and a sequence of "take the square" occupations that spread from Madrid to New York and Oakland and, later on, to Rio, Istanbul, and Kyiv constituted a new wave of social struggles. These tumults displayed the new class composition of digital capitalism: the layers of surplus populations (dramatized in the suicide of Mohamed Bouazizi, the impoverished street vendor whose death catalyzed popular revolt in Tunisia); the youths in edufactories and "the graduate student without a job" (Mason 2012); the neoindustrial proletarians leaping from dormitories in Foxconn plants; and the myriad precarious, low-wage workers who filled squares from Cairo to New York. In many different, and specific to their, locales, the revolts could nonetheless be traced to common threads of indignation at oligarchy, corruption, inequality, and precarity.

No aspect of these movements attracted more attention than the protestors' use of social media, mobile communication, and digital networks. Reportage of Facebook, Twitter, or YouTube "revolutions" has certainly fetishized this activity (Dyer-Witheford 2015). Nonetheless, the 2011 unrests did occur within global populations for whom the use of networks, computers, and, especially, mobile phones was becoming ever more widespread, and who put them to use in rebellious demonstrations, riots, and assemblies. Observers such as Paolo Gerbaudo (2012) and Linda Herrera (2014) have convincingly described the importance of social media "take the square" occupations in Cairo, Madrid, New York, and elsewhere, in terms of the issuing of calls to occupation, logistical organization, circulation of news, and links into mainstream media coverage.

The 2011 struggles also involved major leaks and hacks explicitly regarded by both the perpetrators and enraged state authorities as a form subversive cyberwar (Greenberg 2012). These included the disclosures of WikiLeaks (Assange 2012) and its battles against the retaliatory actions of the U.S. state; the DDoS counterstrikes in support of WikiLeaks by Anonymous (Coleman 2015); and other interventions, such as those of RedHack in Turkey in support of the Taksim Square occupation in Istanbul. The protagonists included defectors from the now digitized military–industrial complex, such as Chelsea Manning and, later, Edward Snowden; veterans of hacker subcultures, such as Assange; and a younger generation of dissidents familiar with chat rooms, digital pranking, music piracy, and ready-made hacking tools, such as those used by Anonymous (Deterritorial Support Group 2012). The groups involved in leaking and hacking sometimes gave direct support to street protests, as Anonymous did to the uprising in Tunisia (Jordan 2015). More generally, there was a strong resonance between hacker activities and popular outrage against unaccountable, venal power; Anonymous's masks appeared on streets and squares from Cairo to New York to Istanbul, becoming the most general icon of revolt.

The participants in this cycle of struggles were played upon from above by the agencies of interstate cyberwar. As we have seen, Russia's "Gerasimov doctrine" holds that this was the case with U.S. support of some uprisings in the Arab Spring (i.e., Libya, Syria) and in Ukraine. It can be reciprocally suspected that Occupy Wall Street protestors were targeted by propaganda agencies such as Russia Today and that (as we discuss in chapter 3) WikiLeaks may later have been strategically fed information about the 2016 U.S. presidential election by Russian operatives. Given the development of interstate cyberwar capacities that we have already detailed, this sort of activity has to be considered extremely likely, even though details of specific events are often opaque. But it is also our assertion that uprisings such as those in the aftermath of the 2008 crash cannot be reduced to effects of such interventions. On the contrary, they arise from the intrinsic, organic conflicts and contradictions of global capitalism, of which both interstate and interclass conflict are a part.

However, if these struggles waged cyberwar from below, they resulted in many tumults but few victories. Many Occupy movements failed to grow beyond the squares they filled and were eventually extinguished. Even where they toppled governments, as in Egypt and Ukraine, secular and left components were later electorally eclipsed by fundamentalist and nationalist elements. The unrests had an "up like a rocket, down like a stick" quality that relates to digital platforms (Plotke 2012). The speed, scale, and contagion of social media rapidly assembled heterogeneous crowds with shared grievances but few common goals; circulated news quicker than alliances could form; catalyzed the rapid start-up of struggles but also their ephemeral fragmentation; and gave social militancy brilliant visibility but also subjected it to merciless surveillance (Pietrzyk 2010; Wolfson 2014). After riots and occupations, police "cyber-crackdowns" used video surveillance, mobile phone records, and social media traces to make arrests (Comninos 2011). Hackers and leakers paid an especially high price; Manning was imprisoned; Assange was trapped; Anonymous was prosecuted; Snowden is in exile.

This sobering trajectory is expressed in both the narrative and title of Zeynep Tufekci's (2017) *Twitter and Tear Gas: The Power and Fragility of Networked Protest*, one of the most recent evaluations of social movements in the era of cyberwar. Drawing heavily on the experiences of Turkish digital activists under the authoritarian Erdoğan regime, but ranging more widely, Tufekci progresses from early chapters suffused with optimism about contagious, cascading revolt to an increasingly somber documentation of the methods by which states recapture networked dominance—not so much by shutting down digital communication, as the Mubarak government attempted in Egypt in 2011 (thereby probably stoking opposition), as by assembling digital militias to colonize and disrupt social media; implanting sabotage and surveillance malware on rebel networks and devices; generating torrential digital propaganda not so much to persuade as to confuse and paralyze; and stonewalling implacably to wait out uprisings until they fissioned internally. Facing this bleak accounting, in the final chapter, she summons up political confidence to remind readers of the still open and renewable horizons of progressive digital activism. This is a fair and necessary point, and one we will take up later. But in the wake of the postcrash rebellions, even worse situations than those Tufekci chronicles awaited many participants. As uprisings in Ukraine, Libya, and Syria spiraled into armed conflicts, foreign military interventions, and civil wars, entire populations found themselves on electronic battlefields that were at once brutally material and inescapably virtual. At this point, the uprisings of 2010–14 entered onto a terrain that already been mapped by another kind of insurgence—one that actualized the Marxist concept of "revolutionary war" in the digital realm, but in a terrible, perverse form.

REVOLUTIONARY CYBERWAR

The concept of revolutionary war was developed by Lenin, Trotsky, Mao, Giap, Guevara, and Cabral. Today the bases from which such wars were supported—the USSR and Maoist China—have collapsed, and the few remaining Marxist insurgencies have been mostly defeated. However, not only has "people's war" been crushed; its doctrine has been usurped by a movement antithetical to its aims—a theocratic, reactionary fundamentalism that has successfully mobilized some of global capitalism's most abandoned and insulted populations in armed struggles that have raged for over two decades. Militant Islamic fundamentalism is a legacy of the Cold War; al-Qaeda was incubated by the U.S. arming of Afghan mujahideen fighting the Soviet invasion of Afghanistan. It is less acknowledged that the theoreticians of armed jihad have clearly studied Lenin, a point that the secular left has been reluctant to engage. The collective Retort (2005) has made one of the few clear-eyed assessments of how far "revolutionary Islam" draws on a Marxist lexicon, adopting both much of the language of third world anti-imperialism and the concept of a vanguard party leading global struggle. And, Retort points out, armed jihad embraces the possibilities modern technologies afford for the waging of guerrilla war.

If it was Zapatistas who first alarmed the U.S. national security establishment about cyberwar from below, it was a very different insurgency that would make them aware of the full cost it could extract, one whose voice was not that of Subcommandante Marcos but that of Osama bin

Laden. The destruction of the World Trade Center in 2001 was certainly a "kinetic" attack—but a digitally facilitated one. As Retort (2005, 153) points out, during the period that al-Qaeda was building its strength, "its essential infrastructure [was] a hard drive and its organizational pathways built around a computer file." Translated from the Arabic, *al-qaeda* means "base"; it has been claimed that this could also be rendered as "database," and although this suggestion may not be linguistically convincing, it does convey the importance of digital networks to al-Qaeda's early operations. Members of the Hamburg cell that carried out the 9/11 attack had communicated via encrypted email, as had the teams that carried out the earlier bombings of U.S. embassies in Nairobi and Dar es Salaam.

However, armed jihad's use of the internet extended well beyond operational planning. Al-Qaeda set up its first website in 2000. By 2003, a widely distributed list of "Thirty Nine Principles of Jihad," considered an al-Qaeda training manual, explicitly specified the waging of "electronic-jihad" as its thirty-fourth item, giving it "paramount importance" because "the Internet provides an opportunity to reach vast, target audiences and respond swiftly to false allegations" as well as to destroy American, Jewish, and secular websites (Leyden 2003). When YouTube launched in 2005, it was rapidly discovered by al-Qaeda, and the wider pluriform universe of armed jihadism surrounding it, as a means for circulating the speeches of its leaders, the wills and testaments of suicide bombers, and infamous decapitation videos in a roiling and mutating culture of propaganda, inspiration, and recruitment, constantly metamorphosing to evade deletion or disruption by increasingly urgent counterinsurgent cyberoperations (Weimann 2006).

The fear jihadist networking struck in U.S. security circles after 9/11 would become a major driver of U.S. cyberwar preparation. It was the failure of U.S. security agencies to "connect the dots" of signals intelligence, which might have given advance warning of the 9/11 attack, that provided the impetus for the superpowered NSA surveillance projects aimed, not at searching data for a "needle" of critical information, but, in the famous words of director Keith Alexander, at "collecting the whole haystack." In turn, this effort was furthered by intelligence operations in the invasions the United States legitimated in the name of the "war on terror."

According to Shane Harris (2014), the development of NSA surveillance capacities was propelled by the example of General McChrystal's tactics against Sunni resistance in Iraq. Conducted under the slogan "it takes a network to defeat a network," these operations implemented counter-insurgency of "industrial scale" (Cockburn 2015) through the interception and analysis of suspect cell phone traffic, followed by special forces raids, directed from high-tech "war rooms," obliterating or incarcerating the entirety of the social networks identified. The apparent success of these operations encouraged the NSA's deepening attention to the potential yields from archiving the metadata of the U.S. domestic population.

When al-Qaeda survived a shattering attack on its Afghan base, the question of the role digital networks played in its organization became critical to major disputes in the West's security establishments. Marc Sageman (2008), a recognized authority on the topic, proposed that Islamic revolution was shifting toward forms of "leaderless jihad" conducted by distributed, decentralized groups connected only by the internet. In his view, this amounted to a significant devolution into weakened forms of ragtag insurgency declining toward the status of a social nuisance rather than a major security threat. Other experts strongly disagreed, insisting on the continued persistence for Islamic jihadism of geographical headquarters, for whom digital networks served to establish logistical and financial chains capable of directing and supplying major attack (Hoffman 2008). In a third position, others agreed on the emergence of "leaderless jihad" but suggested that the internet gave jihadi "self-organization" serious capacities for recruiting and connecting sympathizers, maintaining morale through propaganda and interpersonal communication, sharing knowledge regarding methods and tactics, and developing "novel responses" to counterterrorism practices (Bousquet 2008).

All parties to this argument could only take cold comfort from the eruption in 2014 of the Islamic State in Syria and Iraq (ISIS). One observer of ISIS terms it "the digital caliphate" and suggests that "without digital technologies it is highly unlikely that Islamic State would have come into existence, let alone been able to survive and expand" (Atwan 2015, ix). ISIS combined seizure of a territorial base—at its peak, some two hundred thousand square kilometers—with promotion of geographically distributed self-organizing jihad. Computers and networks were crucial to both. While ISIS troops often avoided electronic communication to evade surveillance, resorting to message runners, they disseminated atrocity pictures via Twitter and other viral media to terrify and demoralize opponents (Brooking and Emerson 2016). In captured territory, ISIS established production centers with skilled staff, high-definition cameras, advanced editing software, and special effects suites to produce the high-quality videos, glorifying the "utopian civil society" of ISIS rule (*Economist* 2015a), celebrating military victories, or displaying the horrifying fate of its opponents, that became a hallmark of its propaganda.

The global distribution of these materials, however, depended on social media networks. ISIS urged followers not only to circulate what they received but to replicate it to mirror sites and other digital archives where it could avoid censorship. At the same time, it encouraged the creation of user-generated content by supporters, fostering a proliferating jihadi online culture whose messages were translated from Arabic not only to English but to Albanian, Bosnian, Filipino, French, German, Italian, Pushtu, Spanish, Urdu, Uighur, and other languages by specialized groups (*Economist* 2016b). A study of a single week's output by ISIS in 2015 found 123 media releases in six languages, 24 of them videos; in the same year, overt ISIS sympathizers held more than fifty thousand Twitter accounts, many across the Middle East but also with concentrations in the United States and Britain (*Economist* 2015b). This means every one of ISIS's messages can be swiftly magnified (*Economist* 2015b). Mastery of anonymization techniques—virtual private networks, Tor, and encryption—allowed ISIS's content dissemination to race ahead of its deletion by hostile state security forces. It is reported that the U.S. State Department's intelligence unit removed forty-five thousand items in 2014 alone and that British police deleted more than a thousand a week, while ISIS social media specialists became targets of drone attacks (Atwan 2015, 25).

ISIS's recruitment of supporters in the West, including international volunteers to fight in Syria and Iraq, was largely virtual. Its foundation was a broad ambience of youth-oriented "cool jihad" internet content (including jihad-themed video game mods). Western security forces would become increasingly concerned about the role of this online culture in inspiring "lone wolf" terrorist attacks. Actual volunteering or recruitment overtures take place via Facebook and Twitter, followed up by instant messenger services such as WhatsApp and Kick and by encrypted Skype conversations. ISIS financial transfers were also enabled online via crypto-currencies, and m-payments to disposable mobile phones, while actual fund-raising was assisted by hacking attacks on credit card databanks of retail stores and ATM machines. Between 2015 and 2017, ISIS produced a magazine, *Kybernetiq,* which was distributed through social media chan-nels like Twitter and *Telegram Magazine*—to instruct would-be jihadists on how to wage cyberwar against the West while avoiding online detection (Cuthbertson 2016).

U.S. cybersecurity experts are skeptical of the ability of ISIS to carry out military-grade attacks. But it did achieve at least one coup when, in 2015, its Cyber Caliphate division compromised the computer networks of the Pentagon's Central Command responsible for Middle Eastern Operations, taking over its Twitter and YouTube accounts and distributing names, ad-dresses, and other details of thousands of military personnel across the web. Following the Charlie Hebdo attacks, the French government attrib-uted a large number of hacker intrusions—particularly DDoS attacks—to ISIS supporters.[15] Although these did not inflict major damage, such low-level attacks may be far more disruptive in regions with very weak cyberse-curity provisions, such as Nigeria, where ISIS allies, such as Boko Haram, operate. ISIS hacking against battlefield opponents using electronic com-munications, such as Kurdish forces, and against the websites of Muslim exiles and immigrants opposing them also appears to have been effective.

In turn, these developments fed into the changing features of U.S. counterinsurgency operations. The Obama regime's determination to escape the political embarrassment, cost, and casualties of "boots on the ground" wars in Afghanistan and Iraq led to the adoption of a new strategy. Large-scale military operations were increasingly delegated to regional allies and proxies. The "sharp end" of U.S. military operations became attacks by cruise missiles, semiautomated armed drones, and highly equipped quasi-cyborg social operations teams, adopting a "way of the knife," but with high-tech blades (Mazzetti 2014). This apparatus merges seamlessly with cyberwar: not only are many of its weapons autonomous or semiautonomous cybernetic robots but the nomination of their targets depends on multilayered systems of digital intelligence, including tracking of mobile phones, social media surveillance, electronic aerial observation, and algorithmically compiled profiles that can identify victims on a probabilistic basis, often with disastrous results (Cockburn 2015).

To acknowledge the reality of armed jihadism and its importance to the unfolding of cyberwar is not to deny the iniquities of the so-called war on terror, the degree to which its prosecution has been used by Western states to justify domestic repression, or the vileness of neofascist incitements to Islamophobia. Nor is suggesting that revolutionary Islam is learned from (or independently repeats) the tropes of Marxist vanguard party and revolutionary war doctrine to approve of it. We trust that it is unnecessary to say more than once that we oppose armed jihadism's violent sectarianism, misogyny, anti-Semitism, murderous martyrology, and theocratic doctrines. To discuss revolutionary Islam's cyberwars is, however, to recognize how fundamentalism and neoliberalism are locked in a mutual "dialectic of disaster" in which terrorist attacks "motivate and generate immense remilitarization of the state and its surveillance capabilities all over the world, and . . . unleash new and deadly interventions abroad, which are equally likely to motivate and to fuel new forms of mass hatred and anti-Western resistance" (Jameson 2002, 304). It is thus also to recognize how large a problem jihadist appropriation of the practices and theories of popular armed struggle presents to the reconstitution of revolutionary secular movements against digital capitalism.

MEETING THE MINOTAUR

The cyberwar labyrinth is made up of crisscrossing paths of technological development, state conflicts, class struggle, and perverse revolutionary wars. But at its center, where these routes intersect, there coalesces a composite monster of superimposed digital conflict, a cyberwar Minotaur. As the wave of 2011–14 activism ebbed, the overlapping dimensions of cyberwar were increasingly apparent.

The hacktivism of social movements, from both the left and the right,

was becoming deeply implicated in intercapitalist conflicts. National security states, acting from above, were proving adept at prompting, penetrat-
ing, and manipulating online populisms rising from below. The collision
of U.S. "soft power" projection with Russia's "information warfare" had
already been building for years. It was further intensified when a wave of
anti-Putin protests at the time of Russia's 2011 elections was given explicit
encouragement by Hillary Clinton. In 2013–14, the Maidan uprising in
Ukraine, an epic, months-long "Facebook revolution" combining "take
the square" occupation and digital mobilization, challenged the kleptocratic, pro-Russian Yanukovych government of Ukraine. The uprising
had wide support in Kyiv and western Ukraine, but in the street, fighting
against security forces that culminated in the flight of Yanukovych proved
heavily reliant on neofascist and ultranationalist groups. The Russian
response included not only annexation of Crimea and military support
for a separatist revolt in eastern Ukraine but also a massive campaign of
digital disinformation, espionage, website defacement, and critical infrastructure attacks (Aliaksandrau 2014; Coker and Sonne 2015; Darczewska
2014a, 2014b; Farmer 2014; Jones 2014; Maurer and Scott 2014). The new
Ukrainian government emerged, to the intense disappointment of many
of the Maidan protestors and their supporters, as dominated by oligarchs,
militarily dependent on right-wing militias, an economic client to Western
neoliberalism, and chronically weakened by internal corruption. It replied
in kind to the Russian interventions, celebrating the exploits of a "cyborg"
soldiery and patriotic hackers (Toler 2014).[16] In the Donbas, as in any war
zone nowadays, civilians and soldiers alike died clutching mobile phones.

In the Middle East, the aftermath of the Arab Spring brought similar
"hybrid" virtual–physical wars driven by the concatenation of domestic
conflicts and international rivalries. In Libya, insurgents against the Gad-
hafi regime vectored NATO air strikes against government forces using
computers, geolocation software, and satellite relays (Scott-Railton 2013).
In Syria, the confrontation of the state with what was initially an unarmed
protest movement spiraled toward civil war and foreign intervention,
with the Assad government eventually supported by Russian warplanes
and Iranian forces in the field and the arming and training of rebels by
the United States. As the conflict became what some suggest is the most

digitally mediated war in history, the making (and faking) of YouTube videos of atrocity in staggering volumes became the virtual face of fratricide. It was also in Syria that the rapid education of authoritarian states in how to digitally respond to Facebook revolutions became fully apparent. The Assad regime did not generally shut down internet connections. Rather, if anything, it liberalized and encouraged social media use as the civil war intensified, to better digitally surveil, identify, and entrap activists. It also mobilized its own Electronic Syrian Army to hack opponents' sites, distribute malware, and harass overseas critics (Deibert 2013). The equation was quickly complicated by the appearance of digitally adept jihadist forces. In a move that exemplifies the unpredictability and opacity of cyberwar alignments, Anonymous (many of whose outlaw activists were languishing in U.S. jails) in 2015 declared cyberwar on ISIS, reportedly disabling websites and social media accounts and thereby entering into a de facto alliance with the Pentagon.

Meanwhile, any idea that digital activism had an inherent "progressive" left, or even liberal tendency, was rudely contradicted. In the United States and Europe, the failure of Occupy movements to achieve political breakthroughs opened a space for far-right forces to capitalize on the immiseration caused by neoliberal globalization. Especially in deindustrialized areas struck by the loss of manufacturing jobs, the linkage of unemployment to immigration issues, racism, and hostility to elite establishments provided a powerful fuel for neofascist political movements fully capable of using networks to advance their programs. Another "firebrand wave of digital activism" began to blaze, but now blown by a wind from the right.

These multiple cyberwar dynamics converged on the U.S. "election hacking" scandal of 2016. Scores of countries have been subject to foreign electoral interventions, very often at the hands of U.S. intelligence agencies and "soft power" manipulators,[17] but the eruption of such interference in the imperial heartland was extraordinarily newsworthy. As we write, the U.S. Department of Justice (2018a) has indicted thirteen Russians and three Russian organization, including the notorious Internet Research Agency "troll farm," for various forms of interference in the U.S. election aimed at assisting Donald Trump (Apuzzo and LaFraniere 2018). The activities of the charged operatives had allegedly been ongoing since 2014.[18] It can

be speculated that the campaign was a response to Clinton's encouragement of anti-Putin activists and to subsequent U.S. sanctioning of Russian
elites in the aftermath of the occupation of Crimea. The Russian foreign
minister dismisses these charges as "blather" (BBC 2018d). The issue of the
Trump campaign's collusion hangs in the air. We think there was Russian
digital "meddling" in the U.S. election but also that its significance has
often been misreported. The reality of Russian digital election interference has to be not only disentangled from its imaginary representations
but also situated in relation to several other "information operations"
(Weedon, Nuland, and Stamos 2017; Stamos 2017), all of which, in their
nested interaction, contributed to the Trump victory, and all in different
ways manifesting the class and geopolitical vectors of cyberwar today.

The first of these is the Trump election campaign's own "military
grade data-driven psychometric micro-targeting" of voters (Albright
2016b). While digital electioneering is nothing new, it is generally agreed
that its deployment by Trump's information manager Brad Parscale was
exceptionally adept and far outdid that of the Clinton campaign. One
reason for this success was a lucid understanding that "when you spend
$100 million on social media, it comes with help"; that is to say, large
expenditures for social media advertising buy the active participation
of the recipient platforms in refining the targeting and distribution of
this material, so that Facebook staff were actively assisting the Trump
campaign (Loizos 2017).[19] Another factor, however, was the services of
Cambridge Analytica, a data-mining and voter-profiling company paid $5.9
million by the Trump campaign to help target messages on hot-button
issues to voters in electorally important locations (Kirschgaessner 2017).
Cambridge Analytica was a subsidiary of Strategic Communication Laboratories, a U.K.-based behavioral research and strategic communication
company with a long record of both electoral and wartime activities in the
developing world that fall squarely in the domain of what are militarily
known as "psyops" and reflect lessons learned in the occupation of Iraq
and other "dirty" wars (Briant 2015). In 2018, it was revealed by former
Cambridge Analytica employee Christopher Wylie that the firm had
purchased from a third-party researcher data gathered from a Facebook
personality-test app. The app, deployed in 2014, not only collected the quiz

results supplied by the 305,000 people who installed it but also gathered the Facebook data of their unknowing friends. It thus harvested information on what Facebook eventually admitted to be up to 87 million people, the majority in the United States (Lapowsky 2018). Cambridge Analytica (which, in the aftermath of the scandal, facing several criminal investigations, closed) claims it did not use these data during its interventions in the U.S. presidential election. However, Wylie asserts that the data were the foundation of the company's entire operation and provided a basis for testing and targeting bespoke anti-Clinton messages (Cadwalladr 2018). While the effectiveness of the "psychometric" influencing techniques used by Cambridge Analytica is hotly debated by political scientists, the episode raises the possibility that an offshoot of the Anglo-American cyberwar complex may have been as or more important to the Trump campaign than any Russian agency.

A second, crucial factor in the election was the "ecosystem" of digital sites connecting the concoction of right libertarians, white supremacists, men's rights groups, meme provocateurs, and disaffected young men making up the "alt-right" (Benkler et al. 2017). Its origins lie partly in in the Gamergate mobilization of "traditionalist" male video and computer gamers trolling feminist critics of digital play's culture of "militarized masculinity" (Kline, Dyer-Witheford and de Peuter 2003), a reactionary reappropriation of the milieu from which Anonymous emerged (Nagle 2017). This was an inspiration for Steve Bannon's positioning of Breitbart News as a platform central to a wider array of far-right sites, such as Daily Caller, the Gateway Pundit, the Washington Examiner, Infowars, Conservative Treehouse, and Truthfeed, with strong links to conservative mainstream media like Fox News (Benkler et al. 2017). However, as Jonathan Albright (2016a) has pointed out, these prominent hubs are in turn connected to a far wider array of several hundred alt-right sites, which he characterizes as a "micro-propaganda machine . . . a vast satellite system of rightwing news and propaganda" surrounding the mainstream media system. This can be seen as a neofascist riposte to left hacktivism in the class cyberwar being fought out over the economic ruins left by the crash of 2008. It provided the matrix within which "fake news" stories supporting the Trump campaign were generated. It was also a factory for sophisticated gaming of search and social media algorithms to give these stories network prominence and for the gathering of data about users that were fed back into the Trump campaign's electoral profiling. Russian digital interventions in the election would have found little traction without this preexisting nativist far-right ambience.

A third element was the purely mercantile generation of pro-Trump news and fake news. Profit as well as ideology was a force in alt-right networks, but piggybacked on those efforts were other cynical commercial ventures. The most famous is the digital industry created in the Macedonian city of Velles. An uncanny distant analog of the U.S. Rust Belt regions critical to Trump's victory, Velles had once been a relatively prosperous industrial Yugoslavian city. The end of the Cold War and the Balkan civil wars of 1991–2001, accompanied by U.S. and NATO military interventions, saw the breakup of Yugoslavia and the formation of independent successor states, leaving the city an impoverished, deindustrialized shell. In 2016, the propagation of several hundred advertising-attracting pro-Trump websites briefly allowed some young men in Velles a sudden, short-lived rise to prosperity (Kirby 2016; Subramanian 2017). They seem to have been utterly indifferent to U.S. politics. When this activity was reported on by U.S. media, comments showed that some Central European readers regarded any ill effects on the U.S. body politic as suitable revenge for the West's role in the disintegration of Yugoslavia (Harris 2017).

It is as a yet further, fourth element in the ensemble of forces converging around the Trump campaign's digital offensive that the intervention of the Russian operatives has to be evaluated. This intervention had several elements. One was probably the hacking and dissemination the DNC emails. Another was the generation of a variety of spurious news stories, some directly supporting Trump or denigrating Clinton, others more diffusely intended to chaotically exacerbate tensions around issues of race, sexuality, and religion. A third was the placement of similarly purposed advertisements on Facebook and other platforms. A fourth were non- or partially virtual activities, such as the building of props (including a cage for an imprisoned "Clinton" figure at Trump rallies). At U.S. Senate Intelligence Committee hearings, the digital corporations across whose platforms this campaign was waged provided some indications of the scale of the digital operations. Facebook initially reported that some 126 million Americans had seen at least one of 3,300 ads placed from about 470 suspect accounts, for a cost of $100,000, a number it later raised to 150 million when Instagram postings were included (White 2017); Twitter identified 36,746 Russian-linked bots that tweeted a total of 1.4 million times in the two months before the election, tweets which it estimates were viewed 288 million times (Reynolds 2017); on YouTube, Google found 1,108 videos it linked to the Russian campaign that netted 309,000 views (Reynolds 2017).

The Russian digital election "meddling" was small in terms of the total campaigning and advertising costs of the contending candidates, a point often made to dismiss its importance. It is also the case that its gradual discovery has been accompanied by misreporting and interested misrepresentation; spurious stories about Russian hacking of the U.S. electricity grid; inadequately supported claims of interference in the British Brexit campaign and European elections; invented statements ascribed to Vladimir Putin; long, indiscriminate lists of purported front organizations for Russian propaganda—all this, and much more sloppy reporting from a commercialized American mainstream news–entertainment system where serious investigative journalism is a fading art become "fake news about fake news" and "propaganda about propaganda" (Greenwald 2017a, 2017c, 2018; Chen 2016). The Democratic Party establishment has an interest in attributing its party's defeat to Russian hacking rather than acknowledging the failures of the Clinton campaign or, more importantly, its complicity with the politics of inequality.

The underlying causes of the capture of the presidency of the world's paramount military power by a far-right demagogue lie in the combination of chronic racism and the Rust Belt consequences of deindustrializing global capital. In that context, however, Russian information warfare was an efficient and low-cost but high-circulation and often carefully aimed strategy for intervening in U.S. politics. It was only one part of several cyberwar dynamics in play in the Trump campaign. It is not only in Russian "election hacking" but also in British psychological warfare techniques adapted for profiling and targeting electoral data subjects; in homebrew American networked neofascism with good search engine optimization; in bizarre Cold War blowback from a Rust Belt–like Macedonian city ruined in the defeat of state socialism; and in the dependence of all of these on the user-generated content, high-message amplification, low-curation platform capitalism of Twitter, Facebook, and Google that we see the full and unruly force of cyberwar's geopolitical and class processes.

Cyberwar thus requires a very complex diagram. What we might call the horizontal axis of interstate conflicts has to be seen in relation to a vertical axis that pits these same rival states against a seething mass of (often contradictory) digitized resistance movements (many of which are supported or co-opted by interstate rivalries in various proxy wars), with both horizontal and vertical dynamics arising from, and feeding into, capitalism's most recent technological revolution. These vectors interact to create a series of zigzagging, diagonal, or swerving alignment patterns whose near-chaotic tendencies are enhanced by the decentralization, opacity, and velocity of networked interaction.

At the end of his reflections on war, Balibar (2002, 15) suggests that in "the early twenty-first century," the complexity of war seems to exceed the measure of Marxist theory and its "dialectical terminology":

> "New Wars," combining sophisticated technologies with "archaic" savagery, external interventions with "civil" or endogenous antago-nisms, are everywhere in the global world around us. They seem to be reviving a "Hobbesian" pattern of *war of all against all* rather than a Marxian primacy of class determinism.

There is no doubt that the scene of cyberwar, with its overdetermined brew of clandestine arcane technologies, postsocialist intercapitalist con-tests, and often reactionary popular uprisings defies all easy teleology. However, we think it is important that the recent rise of cyberwar occurs in the wake of a great capitalist crisis—the financial meltdown of 2008 and its subsequent economic reverberations. This crisis was in part the result of capital's deployment of the computer and network technolo-gies incubated earlier in interstate wars and cold wars as instruments to create a low-wage, precarious global workforce and massively accelerate financial speculation. When this concoction exploded, it ignited a series of global conflagrations, burning fast and slow.

In their description of such capitalist crises, Marx and Engels ([1848] 1964) wrote of how "society suddenly finds itself put back into a state of momentary barbarism," thrown into a state that seems like "universal war." Such "universal war" arises, paradoxically, because "there is too much civilization"; the "forces of production" have become "too powerful" for the "narrow conditions of bourgeois society"—the attempt to channel them toward the priorities of commodification and profit brings contradictory and self-confounding consequences. "And how," Marx and Engels ask, "does bourgeoisie society get over these crises?" "On the one hand," they answer, by the "enforced destruction of a mass of productive force," and on the other, by "the conquest of new markets." Although this passage has been subject to considerable exegesis, one point perhaps not adequately explained is how the "enforced destruction of productive forces" happens: it is often described simply as the consequence of the bankruptcies of failing firms. We, however, suggest that one agency of such destruction is the heighted class and interstate antagonisms elicited by crisis, conflicts that, combined with the drive for "the conquest of new markets" and consequent great-power collisions, create conditions of extreme social volatility.

Read in this way, the passage is perhaps the closest Marx came to articulating a general theory of capitalism's war tendencies. It is social antagonisms that drive toward the annihilation of "too much civilization" as productive forces are turned into instruments for their own destruction—as, today, a vast volume of internet traffic is turned back on itself in massive system-threatening DoS attacks or, more generally, means of communication are weaponized in cyberwar. To recognize these conditions is, however, by no means to resolve the problems a war, or a cyberwar, of "all against all" poses for emancipatory struggles, problems that—and here we agree with Balibar—seem in many ways deeply intractable. We return to this question in the final chapter, but before that, it is necessary to examine more deeply the psychosocial pathologies of cyberwar.

2 Cyberwar's Subjects

INTERPELLATIONS AND ENTICEMENTS

In wars in Syria, Ukraine, and elsewhere, it is common for soldiers to be contacted on their mobile phones by the enemy, with messages either seductive ("I'll send you my photo," "When is your birthday?"[1]), to inveigle the release of tactical intelligence, or derogatory ("You're nothing but meat to your commanders!" "Your body will be found when snow melts"[2]), to demoralize and terrorize. These messages are far from state-of-the-art digital influencing mechanisms, but, bold and crude, they combine the atavistic directness of ideological recruitment and intimidation with the newest technological means of individual subjectification. Yet, extreme as the conditions of these battlefield conversations are, they exemplify a far broader condition of the denizens of states, quasi-states, and would-be-states involved in conflicts waged across digital networks traversed by campaigns of propaganda, persuasion, and surveillance.

To analyze these conditions, we take as our point of departure the Marxist theorist Louis Althusser's (1971) appropriation of Jacques Lacan's ([1969–70] 2007) early work on the imaginary and symbolic orders. Althusser proposed that in capitalism, ideological state apparatuses (ISAs) generate social identity by "interpellating" or "hailing" subjects to engage them in an imaginary relation with the reality of their exploited condition. This proposition has informed a rich line of media studies analysis. However, recently, a number of theorists—Slavoj Žižek (1989, 1997a), Teresa Brennan (2004), Jodi Dean (2008, 2010), Samo Tomšič (2015), and others—have developed and revised this Marxist–Lacanian position to adapt it to the conditions of digital capitalism. In particular, they have emphasized the elements in Lacan's later thought that elaborate the connection between pleasure (however painful or pain inflicting) and a capitalist discourse that entices ever-unsatisfied subjects to labor, in various forms, in search of an enjoyment (*jouissance*) that is always frustrated by the operations of the very social order that sets it in motion (Lacan 2007, 80–81).

We pursue this line of analysis to investigate cyberwar as a new, personalized, yet far-reaching and border-perforating mobilization and discipline of populations. The sometimes-intentional but often-unwitting labor force of cyberwar, network users are interpellated as subjects by world-market states and would-be states to facilitate antagonisms, form alliances, and engage in cyberbattles in various ways, from DDoS attacks, hacking, and trolling to sharing, liking, or commenting. It is, however, an index of the mounting technological intensity of capital (in Marxist terms, its deepening "organic composition") that this process is increasingly automated, whether through the use of software bots to entice or harass or by the enlistment of the user's computer as weapon component. In these conditions, whatever new identities and alignments are proffered, cyberwar's subjects are also immediately rendered obscure and incomplete. The difficulty of attributing and verifying digital warfare operations conducted at internet speeds by machinic agents means that complicit user populations never know for sure what is going on, or what part they play within it, a disoriented condition that, however, only further encourages extreme compensatory identity assertions, virulent conspiracy theories, the consolidation of internet echo chambers, yet greater automation of surveillance to halt real or imagined automated intrusions and infiltration, and yet greater misrecognition by the subjects of cyberwar of their own exploitation and vulnerability.

APPARATUS, CYBERSPACE, NOMAD

When digital networks first diffused from their military point of origin out through society, they were widely greeted as an antidote to indoctrinating powers of state and corporations, a technology of freedom.

Even for those who never read Althusser, in spirit at least, the slogan was "NPCs [networked personal computers] are the antithesis of the ISAs." Many today are reluctant to give up this hope that digital networks are inherently liberatory, a position that has two variants, liberal and radical.

The liberal, originally libertarian, version was that of the North American digital counterculture (Markoff 2005; Turner 2006). Orbiting initially around points such as the early electronic bulletin boards of The WELL (The Whole Earth 'Lectronic Link), *Wired* magazine, and Stewart Brand's *Whole Earth Catalog,* it posited "cyberspace" as an autonomous realm beyond the reach of the state (Barlow 1996), overlooking, at first, and then dismissing, corporate surveillance that has been long intertwined with government surveillance in one powerful complex (Glaser 2018). Those who participated in this perspective were, of course, aware of the military origins of the internet; many had been involved in the anti–Vietnam War movement. But they believed networks had escaped the purview of the Pentagon to become a domain for new virtual identities and communities, a realm of psychological self-actualization. Strongly associated with small hacker start-up companies, this perspective was scornful of corporate suits and legacy media but by no means necessarily anticapitalist. On the contrary, networks often appeared as the fulfillment of the horizontalism of the market's invisible hand, stripped from the vertical accretions of state and monopolist power. Such network libertarianism would eventually be assimilated into neoliberalism, decomposing into a "Californian ideology" (Barbrook and Cameron 1996) that celebrated both the free authenticity of online communication and the Reaganite deregulation that proclaimed the dismantling of the state while funneling giant defense contracts to Silicon Valley. This position didn't so much contradict Althusser as move in a different universe, or at least one separated by an ocean, a continent, and three hundred years of political history. Such belief in the inherently emancipatory powers of the internet has, in its corporately co-opted form, circulated around the planet and remains an article of faith for many users in the age of Google and Facebook.

The radical account of the internet's counter to the ISAs was far more attuned to the issues addressed by Althusser. It in part issued from much the same European left milieu—at least if we accept that one of its basic texts was the extraordinary "nomadological" writings of philosopher Gilles Deleuze and antipsychiatrist Félix Guattari (1986). Their immediate point of reference is the ancient warfare of the nomadic steppe warriors made famous by Genghis Khan. Deleuze and Guattari characterize the culture of these mobile, decentralized nomad warriors as a "war machine"—but one independent from and opposed to any sedentary state formation and, hence, paradoxically separated from any notion of "war" as conventionally understood today. In their account, the fighting aspects of nomadism are subordinated to the transformative processes of "becoming" generated by the new assemblage of humans, horses, and metallurgy brought into being by nomad war. Such "war" is an activity that *precedes* the state and has to be appropriated by it to produce the systematized and rationalized violence we now know by that name.

Though Deleuze and Guattari focus their account of war machines on medieval nomads, they conclude with a reference to the late twentieth century. This, they admit, is an era that the state "war machine" dominates but, nonetheless, still contains possibilities for nomadic reappropriations of its technologies, creating new possibilities for dissenting, transgressive becoming. Whether they had in mind digital networks is impossible to tell, although Guattari's involvement with alternative media makes it likely: the connection is absolutely explicit in Deleuze's (1992) later, prescient "Postscript on the Societies of Control," which also recognized the rapid state appropriation of such powers. But others, such as Manuel DeLanda (1991), rapidly spelled out the affinity between their account of the decentralized, swarming, and rhizomatic nomadic war machine and packet-switching, many-to-many, digital "meshworks," with a potential to subvert the control of state masters. This idea was taken up by anarchists and autonomists associated with alterglobalism and its independent media centers, hacktivism and electronic civil disobedience, a political ambience that found its fullest articulation in the work of Guattari's friend and political ally Antonio Negri and his concept of an emergent revolutionary "multitude" armed by new technologies (Hardt and Negri 2000, 2017). In such accounts, networks appear not, as in the liberal version, just as an escape from the state but also as counterpower, militantly contesting it.

In the liberal view, the "electronic frontier" is a site where self-determining individuals escape from the tyranny of the state and establish communities with like-minded people whose adoption of virtual identities shrugs off socially imposed masks in a new digital authenticity. The radical view is that of a process that dissolves the individual subject in metamorphoses of decentered "dividuals" with new collective possibilities, drawing lines of flight that are simultaneously lines of fight against commanding power. Both, however, see digital networks undoing an ideological authority that polices entry into the world of symbolic communication. Both reflect a certain historical reality, the moment when digital networks, having spread from the US military complex into civilian use, develop without clear state regulation or model for commercial exploitation, as witnessed by the great dot-com crash of 2000. In this window of indeterminacy, all kinds of networked social experiments and political possibilities did indeed bloom.

The context of cyberwar is, however, very different. It is, for one, a context where capital has (in part by the capture of radical experiments) found a business model adequate to the exploitation of networks in the rationalization of "Web 2.0," with its user-generated content, algorithmically targeted advertising, and big data–sucking surveillance. And it is, as we have seen, one where increasingly, states, quasi-states, and their proxies, working through and with this new digital capitalism, strive to consolidate and exercise their monopolies of violence, one against another. Today both liberal and radical insistence on the antistatist valence of the digital networks mystifies and occludes what is unfolding in a way that can itself be termed ideological.

LEVÉE EN MASSE

The state reabsorption of the digital war machine can be tracked in the eager reading of Deleuze and Guattari's nomadological texts by U.S. defense intellectuals.[3] However, for our purposes, a more relevant historical analogy is that propounded by Audrey Cronin (2006), an American academic counterterrorism expert, in her widely discussed account of cyberwar as the new *levée en masse*. The *levée en masse* was the policy of military conscription adopted by the French state in the aftermath of the

Revolution of 1789 in a desperate attempt to counter the superiority of the professional ancien régime armies sent to crush it. It was unexpectedly successful. Numerically superior and (sometimes) fervent citizen-soldiers shattered the ranks of their opponents at Valmy and other battlefields. While a matter of state policy, *levée en masse,* as Cronin emphasizes, harnessed revolutionary popular enthusiasm; in French, *levée* signifies not merely a governmental imposition, such as a tax, but also an uprising. As Cronin (2006, 79) points out, it drew on the new means of transportation (good roads) and communication (cheap printing) of a nascent mercantile capitalism to recruit its soldiers and get them to the front. With the collapse of censorship, the circulation of revolutionary tracts, songs such as the "Marseillaise," and images, such as pictures of the storming of the Bastille, ensured that "the French populace was reached, radicalized, educated, and organized so as to save the revolution and participate in its wars."

Then Cronin draws her parallel with contemporary cyberwars. Pointing to the "global spread of Islamist-inspired terrorist attacks . . . the rapid evolution of insurgent tactics in Iraq, the riots in France, and well beyond," she argues that this is "the 21st century's *levée en masse,* a mass networked mobilization" (77). With her eyes on al-Qaeda and digital jihadism, she suggests that "democratization of communications, an increase in public access, a sharp reduction in cost, a growth in frequency, and an exploitation of images to construct a mobilizing narrative" are all working in favor of "so far uncontrollable insurgency" (81). Like the *levée en masse,* the evolving character of communications is "altering the patterns of popular mobilization, including both the means of participation and the ends for which wars are fought" (84–85). "Today's mobilization," she continues, "may not be producing masses of soldiers, sweeping across the European continent, but it is effecting an underground uprising whose remarkable effects are being played out on the battlefield every day" (85). "Cyber-mobilization" is bringing about broad "social, ideological, and political changes"; successfully harnessing these elements is "the key to advantage in future war." In the alarm modality of post 9/11 U.S. military thought, she declares that "the information age is having a transformative effect on the broad evolution of conflict, and we are missing it" (87).

Cronin's analysis came in the early days of the U.S. "war on terror." We know now what form the "counter-mobilization" (87) that she urged against Islamic militancy would take: mass surveillance, drone strikes, and Special Forces ops. Other aspects of cyberwar, such as its state versus state modalities, have subsequently come to the fore. However, her article retains its relevance because of the way it articulates a project of state recapture of an initially "revolutionary" force—the emergence of networked populations.[4] There is (and was) a deep political ambivalence in the concept of the *levée en masse,* one caught perfectly in the subtitle of Jean-Paul Bertaud's (1988) study *The Army of the French Revolution: From Citizen Soldiers to Instruments of Power.* The *levée* was indeed radical: it saved the Revolution. It then, however, became the basis for the triumphs of Napoleon's armies, in a project of conquest that founded a new and emphatically imperial power. It was also emulated and adapted by Napoleon's reactionary adversaries, providing a general war-making model for nineteenth-century, and then twentieth-century, great powers. Insofar as national conscription formed the basis for total war, a line connects Valmy to Verdun.

For Cronin, cyberwar is a matter of "mobilization." Geoffrey Winthrop-Young (2011, 134–35), discussing the military writings of Kittler, describes the category of mobilization well and makes the connection to issues of subjectivity:

> Mobilization erodes the boundaries between war and peace because it takes place in both; it erodes the boundary between the military and civilian population because it affects one as much as the other; and it erodes the distinction between material hardware and psychic software because it deals as much with the optimization of logistics, transport, and technology as with increasing mental preparedness and overall combat readiness. But what kind of human is most equipped (or least under-equipped) to deal with the acceleration and incomprehensibility of modern war? What kind of mind is available to make rapid, on the spot decisions, or even make up new rules when no fiat, no commanding authority, is in sight? What has been programmed to fight with a free will? The modern subject.

It could be argued that cyberwar, a form of highly technocratic warfare, is in some regards the opposite of the *levée en masse,* a type of war that, like nuclear weapons, frees states from their politically problematic dependence on mass armies. However, as we suggested in the previous chapter, this idea of cyberwar simply as a series of hacker-team exploits ignores the wider base of technosocial knowledge and practice on which such feats depend. It also ignores the global networked populations that cyberwar hacking traverses, targets, and exploits.

Hardt and Negri (2000) and others of the post-*operaismo* school have interpreted Marx's (1973) passing reference to a "general intellect" as an allusion to the collective skills, aptitudes, and identities necessary for capitalism's ceaselessly innovative technological development. Much the same logic applies to its concomitant development of technowar, a dynamic in which the "general" in "general intellect" could be taken as referring not only to a collective or communal process but one under military command! If we restore this process to view, the concept of mobilization, including the formation of specific subjectivities for cyberwar, makes sense, although it is a mobilization that, as we will argue later, is as much concerned with activating machines as it is with galvanizing users.

It is a form of mobilization that has novel features. Competing states have always engaged in usually clumsy and ineffective propaganda wars aimed at disaffecting their opponents' populations. On a networked planet, however, it is not just the homeland state that interpellates its subjects with the exceptional intimacy and intensity afforded by digital systems but also the enemy—the adversary state—that can do so. Before we develop these arguments further, we will give two examples of cyberwar mobilization. In both cases, we see how an initial revolutionary or insurgent use of the internet is appropriated by state apparatuses, those that emerge from the initial rebellion *and* those that seek to quell it or initiate new revolts against its outcomes. This dynamic leads to an escalating militarization of networks and the intensifying formation of warring data subjects. The first example involves conflict in Ukraine, the second in Gaza.

In Ukraine, the 2013–14 uprising of the Maidan against the Yanukovych oligarcho-kleptocratic regime, a classic "Facebook revolution," brought to power the government of President Poroshenko. Four years after, at the moment of writing this book, the new president himself is a center of several large scandals—from the disclosure, by the Panama Papers leak, of his secret offshore company, Prime Asset Partners Ltd, in the

British Virgin Islands, set with the goal of tax evasion (Harding 2016), to his lavish, half-million-dollar secret trip to Maldives at the beginning of January 2018 under the name of "Mr. Petro Incognito (Ukraine)," abandoning the country at war (Romanyshyn 2018b), since 2014, against pro-Russian separatists supported by the Russian military in the self-proclaimed Donetsk People's Republic (DPR) and Lugansk People's Republic. This is variously termed a *proxy war* (in the international and some Ukrainian media, emphasizing major Russian assistance to the separatists), *civil war* (by the Russian government and in many Russian media, emphasizing the division of Ukraine's population to deny the presence of the Russian troops), *antiterrorist operation* (ATO) (by the Ukrainian government until February 2018), and *armed aggression* (by the Ukrainian government since February 2018).[5]

Both sides, but especially pro-Russian forces, have made use of digital networks for computational propaganda over social media, digital espionage and hacking, ranging from DDoS attacks to the blackout of a major section of Ukraine's electric grid, launched from a Russian IP address. The Ukrainian side often mirrors the propaganda methods and tactics of the northern neighbor; for example, a Ukraine-based private English-language satellite television channel, Ukraine Today (2014–16), was modeled on Russia Today, insisting on the distinction between "good" and "bad" propaganda. The conflict has been characterized by widespread dissemination of doctored videos and disputed interpretations of critical events (such as the shooting down of Malaysia Airlines flight 17 over Donbas) fought out over social media.[6] There is substantial evidence that reporting on the conflict, in Ukraine, in Russia, and abroad, has been targeted by paid internet "trolling" enterprises located in Russia, tasked with intervening in discussions to depict Ukraine governments as fascist, incompetent, and corrupt, a puppet state of the degenerate, "gay" Western liberal democracies, and Ukrainians as the manipulated victims of imperialist designs to divide them from their Slavic Russian brothers (Sindelar 2014; Chen 2015; Patrikarakos 2017; Soshnikov 2017). This is accompanied by the circulation of faked war videos and fabricated atrocity stories (whose spurious nature, of course, doesn't preclude the existence of real war horrors and atrocities).

For its part, the Ukrainian government has, with the assistance of NATO consultants and advisors, mounted its own information war campaigns. Facebook and other social media platforms are widely used by civil-society organizations supplying Ukrainian troops and militia in the "Anti-Terrorist Operation Zone" (Patrikarakos 2017). Ukrainian hackers, volunteers or government employed, hack information from the CCTV cameras in eastern Ukraine, as well as local insurgents' bank accounts, phone conversations, and emails. Some hackers give information to the Peacemaker (Миротворец) website, indirectly associated with Ukrainian government figures who, in an appropriation of "leak and hack" tactics, call themselves "the friends of Assange" (Миротворец, n.d.). The site offers "information for law enforcement authorities and special services about pro-Russian terrorists, separatists, mercenaries, war criminals, and murderers." Peacemaker published information on nine thousand alleged "terrorists" fighting in Ukraine, calling for an open hunt for these people online and off. Shortly after the site's launch in early 2015, it came to public attention for publishing the personal details of forty-five hundred journalists, complete with phone numbers and emails. The website's creators accused them of collaborating with the "terrorists" because they had received accreditation from the leadership of the DPR.[7]

Both during the Maidan rebellion and the Donbas war, social media usage in Ukraine has split between rival platforms in a way broadly mirroring political divisions. While supporters of the Maidan revolution favored Facebook, and widely used it to coordinate and circulate news of their revolt, the majority of pro-Russian Ukrainians have congregated on VKontakte. Also known as VK (http://vk.com/), VKontakte is a Russian-owned social media platform, founded by Russian entrepreneur Pavel Durov in 2006, that, in its "user interface and functionalities . . . largely resembles Facebook," allowing a user to create a profile (public or private) and then start "'friending' other users" (Gruzd and Tsyganova 2014, 124). In January 2014, after Durov's refusal to turn over data on Ukrainian protesters to the Russian government, the founder was forced to sell his stake in the company to the company controlled by Russian oligarch Alisher Usmanov, a co-owner of Russia's second largest mobile telephone operator, MegaFon, and co-owner of the Mail.ru group, the largest Russian-speaking internet company. Meanwhile, Durov himself had to escape the country. While VK lacks Facebook's global reach, it has more than 100 million users, mostly from former Soviet republics. In Ukraine, VK is—or was—less popular than Google and YouTube but more so than Facebook. On May 2017, however, President Poroshenko announced a ban on VK and several other Russian-owned internet firms: Odnoklassniki, another widely used social network; Mail.ru, one of the country's most popular email services; Yandex, a major search engine; and software from Russian cybersecurity firm Kaspersky. The president described the measures as "an answer to 'massive Russian cyberattacks across the world'"; Ukrainian officials said that the social networks in question were "used to spread Russian propaganda, and that users' data are collected by Russia's secret services" (*Economist* 2017c). The decision was widely protested in Ukraine, but in the month following, Ukrainian Facebook accounts surged by around 2.5 million. After announcement of the ban, Ukraine's presidential administration claimed that its website had come under attack by Russian hackers.

Our second example of digital mobilization comes from the escalating network conflict between Israel and Palestine that has accompanied successive rounds of asymmetrical fighting in Gaza. In his *War in 140 Characters*, David Patrikarakos (2017) suggests that Israeli success in suppressing successive Palestinian intifada (uprisings) from the late 1980s on was based not only on massive military superiority but also on success in controlling framing and narration of the conflict in the international press. Israel's ability to restrict journalist access to the West Bank and Gaza, its cultivation of a relatively narrow and predictable range of foreign news channels, and the widespread sympathy to Israel in liberal public opinion, especially in the United States, all allowed the issue to be represented as a battle against terrorism. However, with the growing Palestinian use of the internet, and then of social media, this dominance is eroded.

From 2000, despite limited access and disrupted electricity supply, the internet became a means for Palestine's diasporic communities, both in Middle Eastern refugee camps and beyond, to share symbols and texts of national identity, a process referred to as an electronic or "cyber intifada" (Aouragh 2003). In the first Gaza War in 2008, designated by Israel as Operation Cast Lead, Israeli Defense Forces (IDF) restricted Western journalists' access to the war zone, resulting in a stream of images, video, and information relayed from the Gaza Strip to media outlets such as Al Jazeera and Al Arabiyya. The proliferation of Web 2.0 services allows the people of Gaza to upload detailed accounts to blogs, Twitter, YouTube, and, notably, Flickr, revealing scenes of "chaotic horror" caused by Israeli air strikes (Taylor 2012). In response, the IDF rapidly assembled a Spokesperson's Unit to produce a social media presence, but, despite persuading YouTube to take down some Palestinian videos, it had limited success. The deterioration of Israel's media position intensified in 2010 with the Mavi Marmara incident. The IDF's interception of a "peace flotilla" of boats breaking the economic blockade on Gaza culminated in an airborne assault on the main vessel, killing nine Turkish activists and wounding many others. Despite the IDF's attempt to block the ship's electronic communication, enough video and audio material was transmitted to convey the violent mayhem of the attack, creating a diplomatic scandal that Israel only slowly recuperated from over the following year.

Confronted with the growing digital capacities of its opponent, the IDF rapidly built its own; the Spokesperson's Unit was filled with young soldiers adept in social media and some four hundred volunteer students recruited from a private Israeli university supported by the Israeli government and the IDF (Rodley 2014). In Operation Pillar of Defense, an eight-day attack on Gaza, the IDF made use of Twitter and liveblogging to give its up-to-date version of events, running a "hyper-pugnacious" online campaign (Shachtman and Beckhusen 2012). Meanwhile, the Palestinians, with the assistance of Anonymous, mounted major DDoS attacks on Israeli websites. With both protagonists acutely aware of the Twitter revolutions of the previous year's Arab Spring, #IsraelUnderFire confronted #GazaUnderAttack.

When, in 2014, Israel launched the fifty-day Operation Protective Edge against Gaza, with sustained aerial bombardment followed by ground invasion, both sides unleashed what Chris Rodley (2014) terms "viral agitprop" disseminated across "crowded, competitive and fast-moving" social media, with videos, infographics, Twitter feeds, "hypermediated" pop-cultural allusions, "clickbait" headlines, and "meta-commentary" on the antagonist's material. This was aimed not so much at the adversary, for antagonisms were solidified and the messaging was conducted over unshared feeds, but, as Rodley puts it, at "winning support from foreign audiences, rearticulating national identity, boosting morale, and . . . neutralizing enemy messaging." Because viral agitprop is distributed in small, real-time bursts, unlike a feature article or television broadcast, a single channel or user is "able to generate a wide range of material performing a diverse range of functions each day of the war" (Rodley 2014). Militarily, Israel won the war, or at least inflicted by far more casualties than it suffered. Politically, in terms of public opinion and propaganda, the outcome was far from clear cut. Since 2014, networked hostilities have continued; the government of Israel has frequently complained to Facebook about pages allegedly containing Palestinian instigations to violence, and many, including those of politicians, bloggers, and journalists, have been taken down; Palestinians claim Facebook does not respond similarly to complaints about pages in Hebrew inciting violence against them (Greenwald 2017b).

With these two examples in hand, as well as others provided in previous chapters, we can perhaps now say that "this is what cyberwar mobilization, or the militarization of general intellect, or the digital *levée en masse,* looks like," so we now proceed to a more theoretical analysis of this condition.

THE CYBERWAR APPARATUS

In Althusser's discussion of ISAs, the famous example is that of a citizen's spontaneous response to an abrupt "interpellation" or "hailing" by police—"Hey, you!"—immediately self-identifying as an obedient subject to the force of law and order. This instance Althusser takes as paradigmatic of how an array of institutions—schools, churches, political parties, trade unions, media—speak to, summon, name, or address individuals in such a way as to induce an apparently autogenerated compliance with the dominant social order. Althusser wrote before the popular adoption of the internet, so if we return to the notion of ISAs, it must necessarily be as a revision, specifying new interpellative institutions operating through what we will for the moment simply call the *cyberwar apparatus.*

This is an apparatus that emerges at a stage in the development of digital networks very different from that of the electronic frontier or the dot-com boom, when notions of digital autonomy and nomadism flourished. The rise of cyberwar is contemporaneous with the ascent of Google and Facebook in the mid-2000s. Cyberwartime is the time of "Web 2.0" and of the reconceptualization of digital media not as publisher but as "platform," managing proprietorial software that offers tools for structured but self-directed network activities and community creation, the monitoring and measurement of which supplies the big data—generated and collected in astounding "volume, variety and velocity," to use a familiar characterization—that are processed and analyzed algorithmically to target the advertisements that are the major revenue source for the great social media and search engine enterprises (Bratton 2016; Srnicek 2016).

In this context, we see some inadequacies in Althusser's concept of interpellation. For one, his theory of ideology, apparatuses, and subjectification was, from the start, too bluntly state-centric. It asserted that the function of ideology was to secure the reproduction of capital, but did not adequately acknowledge how capital in its corporate form itself undertakes this task, as, in the very processes of commodity circulation, it socializes populations to the relations of production and consumption.[8] This, as the Frankfurt School pointed out long ago, is a process especially assumed by the "culture industry." Today, it has hypertrophied into a cycle of 24/7 entertainment, where, Žižek (1997b) says, we are interpellated by a constant exhortation to "fun" and where the disciplinary functions of ideology are interfused with enjoyment (Flisfeder 2018, 42). This is nowhere more so than in the realm of the digital platforms.

And here we can also see that the Althusserian concept of interpellation is too cognitive; what is missing is the body, the *physical body* of the subjectivized individual that is *colonized* within the information economy of capitalism. Today, not only chatting, liking, sharing, gaming, viewing, and other *fun* has become *work* but also walking, breathing, sleeping, and doing nothing, as these activities are submitted to technological surveillance (Cederström and Fleming 2012). By introducing the notion of the corporeal "speaking being," the *parlêtre,* in his later work, Lacan brings forward the physical body that is speaking and laboring in one joint act.

The body as "the site of discursive production," Tomšič (2015, 20) reminds us, "contains two aspects: the production of subjectivity and production of jouissance," which is as much enjoyable as it is exhausting. Just as labor power becomes a commodity in capitalism, so does *jouissance* in "communicative capitalism" (Dean 2009). A striking example is provided by the case of the company Strava, which, in November 2017, published an interactive map showing fitness-tracking activity around the globe (Sly 2018), thereby inadvertently revealing the running paths of military personnel in war zones where only soldiers are likely to wear fit-bits, outlining a suspected CIA base in Mogadishu, Somalia; a Patriot missile system site in Yemen; U.S. Special Operations centers in the Sahel region of Africa; the main Russian airfields in Syria; and other "undeclared facilities throughout Syria belonging to both Russian and NATO forces" (Sly 2018; Scott-Railton 2018). When the communicating subject, the *parlêtre*, is not only at work but also at war, where it speaks, it enjoys itself to death.

In "platform capitalism" (Srnicek 2016), users validate their existence through an assemblage of social media profiles, postings, preferences, inquiries, recommendations, and all the other digital footprints that represent it within a "programmed sociality" (Bucher 2012a). It occupies "a position [that] never allows someone to enter it fully formed," always rendered "comparable and interchangeable through various qualifications and quantifications of behavior and impact" (Bratton 2016, 252) for the purpose of extracting value and, at the same time, incessantly engaged by the scopic regimes of the personalized web. As Geert Lovink (2016) observes, referencing Althusser via Wendy Chun's (2004) earlier "On Software," the user is "interpellated" by social media from the moment of sign-in:

> Before we enter the social media sphere, everyone first fills out a profile and choses [*sic*] a username and password in order to create an account. Minutes later, you're part of the game and you start sharing, creating, playing, as if it has always been like that. The profile is the a priori part and the profiling and targeted advertising cannot operate without it. The platforms present themselves as self-evident. They just are—facilitating our feature-rich lives. Everyone that counts is there. It is through the gate of the profile that we become its subject.

Or as Ganaele Langlois et al. (2009) put it, "commercial Web 2.0 platforms are attractive because they allow us, as users, to explore and build
knowledge and social relations in an intimate, personalized way," while
at the same time paradoxically "narrowing down the field of possibilities" in ways that "favor the formation of specific subject positions." This
is now the battlefield of cyberwar. In the arena of what is all too aptly
dubbed "iWar" (Gertz 2017), the state apparatuses, apparently exceeded
and supplanted by corporate forms of ideological address, abruptly reap-
pear, their interpellations lodged in and relayed by the "likes," "tweets,"
"recommendations," and "follows" at the very heart of digital sociality
and identity formation.

A state's interpellation of its subjects, or of the subjects of another
state, is in part a question of its cyberwar apparatuses' access to or interdiction from specific platforms. Though platforms may be more or less
global or local in scope, they are also national, in terms of ownership and
legal governance and relations to a homeland state security apparatus.
This is a dynamic around which the profit interests of specific blocs of
digital capital and political interests of security state apparatuses mutually
revolve (Google/Facebook/Twitter for the United States; VKontakte/
Yandex for Russia; Weibo/Baidu/Tencent for China). "National" interpellation of state subjects is manifestly in play in the banning by the Ukraine
government of Russian social media and search engines in the midst of
an ongoing war. It may also, however, manifest when military conflict is
merely anticipated, as we saw in chapter 1 in regard to the "Sino-Google
war" (Bratton 2016, 112) and its imbrication in intensifying China–United
States hostilities.

However, cyberwar subjectification is not simply a question of de-
marcating "national" platforms and "national" identities. The profit dynamics of Web 2.0 capital demand both an ever-enlarging user base and
the self-activity (and hence self-revelation) of users. Because of this, it is
possible to construct within social media interpellative micromachines,
that is to say, specific user communities, as a sort of "partisan" presence
inside ostensibly foreign digital territory. Within Facebook, one can form
a nation, or a caliphate, or perhaps even an assembly or commune. Platforms owned by capital of another state, even an adversary state, can be

seeded with subversive practices or used as digital territories across which wars are fought between other competing states, as the Gaza conflict was digitally fought out across Twitter. The truly global social media platforms, owned by U.S. capital, are particularly liable to this process, precisely to the degree that their huge profitability depends on amassing global users. This cyberwar seeding of subjectification involves agents. These are of various kinds: state military public relations agencies, such as the IDF's Spokesperson's Unit (now being widely copied by other militaries); clandestine troll armies, such as those of Russia's Internet Research Agency; U.S. military "sock puppets"; or ISIS militant recruiters. The efforts of such agents may also be articulated with autonomous interpellative processes, such as those of dissenting movements within foreign states or citizen journalists with a spontaneous patriotism for the homeland. Agents may work positively, as "friends"; negatively, harassing enemies; or in contradictory, chaotic directions, generating a paralyzing anomic blur. Each of these comes with its own interpellative processes.

For example, one can contrast the careful cultivation of the subject of an ISIS recruitment process with trolling practices associated with Russian hybrid warfare. In the former process, recruiters "monitor online communities where they believe they can find receptive individuals," sifting through visitors to militant sites. They then create a warm virtual microcommunity around potential recruits, saturating them with messages that, depending on the orientation of the target, may emphasize religious devotion, the appeal of violent action, grievances against racism, religious discrimination, poor economic prospects, or positive depictions of life in militant-held territories. At the same time, they encourage the potential recruit to isolate himself from other contacts. At a certain point, communication shifts from public forums into private and encrypted channels. It is in this phase that options such as emigration to ISIS territory, as a civilian or fighter, or attacks at home are discussed. The cycle is a sustained, modulated "hailing" of an identified data subject as a supporter of the caliphate, a data relation that then translates into corporeal action (Berger 2015a, 2015b).[9]

Cyberwar trolling is an interpellation that reverses the logic of recruiting. An enemy, rather than a friend, is identified, vilified, enraged, exhausted, and metaphorically (and sometimes literally, through "doxing" and the like) destroyed, silenced, intimidated, or forced offline—with the correct subject position established negatively, in contrast to that of the unfortunate and despicable victim. While the term is loosely applied to many forms of online harassment, it is often more specifically used to characterize a strategy of escalating rancorous dissensus, for example, making a provocation that can be bootstrapped into intensifying abuse and insult. In this form, trolling is sometimes analyzed as having a distinct sequence; the lure, the catch, reeling in, and so on. It may be practiced solo or in teams (with one troll reinforcing another or making an apparently innocent conversational setup that can later be exploited) and can involve a number of gambits (such as professing support for a position but then undercutting it with damning "concerns"). Thus, for example, on a nationalist Ukraine internet forum, pro-Russian trolling can ramp up from an apparently measured remark about U.S. support for Ukraine to intensifying attacks on Ukraine's corrupt incapacity for self-governance, Western homosexual degeneracy, ubiquitous fascism, cowardice, and inevitable defeat. The subjectifying address is "if you are like that, you are worthless; you don't want to be like them, the faggots/fascists/CIA dupes, but like us, your brave Slavic brothers" (Szwed 2016).

Though cyberwar apparatuses require agents, they depend on virality (Sampson 2012). A first, minimal sign of a successful digital interpellation is that it generates a "like," a "retweet," a "follow." It is in the nature of the viral process that it becomes difficult to distinguish instigators from followers; it aims at user complicity.[10] Such virality is not just a communication of discursive political positions and arguments. It is a "transmission of affect" (Brennan 2004), a process of emotive contagion. This is why images, particularly images of the horrors and atrocities of war, variously authentic and fabricated, are so important. The IDF uploads video from the helmet-mounted camera of an Israeli soldier in urban combat in Gaza; the fifteen-year-old girl Farrah Baker, one of the most effective of Palestinian citizen journalists, tweets a photo of the night sky above her home blazing with flares in the midst of intense aerial bombardment, accompanied with, "This is in my area. I can't stop crying. I might die tonight #Gaza #GazaUnderAttack #ICC4Israel [International Criminal

Court for Israel]" (Patrikarakos 2017, 30). Both interpellate by eliciting identification with a moment of intense affect—the excitement of combat, the fear of death—that entrains a political sympathy.

There is no guarantee of virality (unless it is artificially boosted by bots); any given interpellation may be wasted, washed away, interrupted, contradicted, and cancelled or co-opted in the babel of social media voices and vanish without a trace. Conversely, various interpellative sequences, with or without common origin, may start to resonate with one another. As we have already suggested, the importance of the clandestine advertisements of the Internet Research Agency around the 2016 U.S. presidential election, with their invocations of threatening immigration, dangerous minorities, Clintonite Satanism, and diffuse but intensified social antagonism, was only as one element in a digital surround-sound interpellation of the U.S. electorate conducted simultaneously from multiple directions—the official Trump campaign, the alt-right's autonomous digital networks, Macedonian for-profit web news producers—that together generated multiplying addresses constructing an aggrieved, white, American–patriotic subject under attack from people of color, Islamic terrorists, political elites, foreign job stealers, and miasmic social anomie. Such compounding and escalating interpellations generate waves of viral affect (Massumi 2015), increasing the probabilities of summoning up a respondent to the proffered subject position.

In 2011, Eli Pariser brought to public attention the phenomenon of "filter bubbles." Since then, the concept has been used to criticize the personalized web for producing a pacifying sense of self-conformity. However, subsequently, we have seen the growing antagonisms within or between these echo chambers. Such antagonisms are exploited by state or insurgent powers that build on the preliminary work of automated segregation performed by commercial algorithms generating consumer profiles. Peter Sloterdijk's (2011) "spheric project" of bubbles, spheres, and foams helps conceptualize these network productions. Between the microspheric bubbles of the intimate and the macrospheric globes of a historicopolitical world lie the "foam worlds" of Amazon, Google, Facebook, Twitter, Weibo, and VKontakte, where "the individual bubbles are not absorbed into a single, integrative hyper-orb . . . , but rather drawn together to form irregular hills," making "what is currently confusedly proclaimed *the* globalization of the world [a] universalized war of foams" (71). This for Sloterdijk is "the modern catastrophe of the round world" (70): the "spheric blasphemy" (69) of antagonized plurality.

In turn, interpellative networked interventions by antagonists spur states to the adoption of homeland censorship and/or surveillance systems, with the aim of blocking or deleting hostile virality and detecting or eliminating its instigating agents. The two elements in this censorship–surveillance ensemble are distinct but related. Censorship implies a monitoring and disciplining of those who attempt to evade it, and surveillance, if known or suspected, results in the internalized discipline of self-censorship. What is in play here, however, is, as John Cheney-Lippold (2017, 169–72) points out, not so much a state's direct address of the subject as its observation and categorization of that same subject, be it as a "good citizen" or "terrorist suspect" or "foreign agent." This disturbs the "clean cut" (Dolar 1993) finality and determinism of Althusser's model of interpellation. It posits a subject that is not talked *to* but talked *about*—a subject of surveillance not usually aware of the identity bestowed on it but that, suspecting surveillance, does its best to avoid suspicion. This subject is not immediately exposed to the "hey you" hail of the police but may become aware of her "composite algorithmic identity" from the changing nature of interpellations, say, at a routine traffic stop, by a guard at a border crossing, or during an airline passport check (Cheney-Lippold 2017, 170).

Panoptic surveillance and virtual mobilization are reciprocally related in complex and contradictory ways. It is the potential for subversive mobilizations that evokes state surveillance, yet states also increasingly themselves mobilize virtual recruiters, troll armies, patriotic hackers, and social media communities against their opponents. This double face of interpellative incitement and surveillant suppression constitutes a feedback loop constantly reinforcing the cyberwar apparatus. This again revises the Althusserian account of subjectification. For with war, we have a *contest* of interpellations, and a subject played upon by and constituted in that collision, a subject not only addressed by the homeland ISAs but also exposed to adversary address and, because of this, then subjected to processes of surveillance and censorship, watching and blocking, that in turn become constitutive of subjectivity.[11]

Because cyberwar interpellations largely depend on for-profit social media platforms, they, at root and regardless of source, reinforce the interpellation of the "user" as a subject of global digital capital. But this interpellation is also and simultaneously the constitution of the subject of particular fractions and blocs of digital capital, aligned against one another in association with antagonistic national security states and would-be states. The interpellative process is thus split by the inescapable contradiction between the one, total capital of a planetary system and the many, competitive, hostile capitals composing that totality. In cyberwar, this agonistic process of interpellation, conducted in the fast flows of social media, intensifies the always unfinished and ineradicably incomplete nature of subjectification, an element of Lacan's thought that Althusser arguably misunderstood or abandoned in his original account of the ISAs.

In summary, cyberwar operates through apparatuses of subjectification that work across platforms and are aimed at the populations of these platforms (populations that can be conceived of as combined human–device assemblages) both at home and abroad. This apparatus employs specific agents (bearing in mind that, as we will discuss later, these agents may be wholly or partially automated) that issue the interpellative call or summons (itself the outcome of long-preceding chains of interpellative subjection). But this call thenceforward depends for its efficacy on networked contagion (whereby if the process "takes," each interpellated subject becomes an interpellator), a process that may go nowhere but whose accumulative outcome can be filter bubbles or echo chambers (the death stars or black holes of cyberwar) of autodisciplining subjects. The preemption and targeting of such digital enemy partisans, real or imagined, become the task of the conjoined surveillance and censorship mechanisms by which each regime's cyberwar apparatus attempts to prevent its own infiltration and disrupt the adversary. This is how cyberwar engages the continuous process of reproducing the "economic, political, juridical and cognitive fiction of the subject" (Tomšič 2015, 6).

Just as Althusser suggested that the ISAs comprised a range of institutions—school, church, media—each of which operated in its own particular register and could be variously articulated with the others, so cyberwar apparatuses are a collocation of institutions and practices—soft power, digital propaganda, troll armies, internet recruiters, surveillance and censorship—that can be variably permutated with one another by any particular state or quasi-state. The old ISAs do not disappear (positing such a world remade *ab novo* by digital networks would be the worst version of ideology) but rather collaborate with, are reshaped and reinforced by, the cyberwar apparatus. Thus, for example, Zeynep Tufekci (2017) notes how under Turkey's authoritarian Erdoğan regime, engaged in the double repression of domestic dissent and war against Kurdish rebellion, a major governmental strategy has been to *discourage* the populace from using social media, widely used by regime critics, disparaging it as a realm of perversion and deceit, aiming instead to retain the people within the orbit of television watching, where the state feels more secure in its control of content.

We are in a zone where "algorithmic ideology" (Mager 2012), "algorithmic governance" (Just and Latzer 2017), or even an "Algorithmic Ideological Apparatus" (Flisfeder 2018) meets with "algorithmic war" (Amoore 2009). However, we do not speak only of ideological effects. Althusser's (1971) original formulation divides the "ideological state apparatus" from the "repressive apparatus," comprising the military and security forces, and emphasizes that, while each has both ideological and repressive components, the former operates "massively and predominantly" by ideology, the latter primarily by violence (145).[12] In the case of cyberwar, we argue that the weaponization of communication means there is a cross-over between the ideological and violent operations; while some activities, such as computational propaganda, fall on one side of the spectrum, and others, for example, a nuclear facility–destroying computer worm, fall on the other, there is an intermediate zone where so-called social engineering and viral propagation are indispensable to both. Althusser reminds us that "there is no such thing as a purely repressive apparatus," because "even though the Ideological State Apparatuses function massively and predominantly by ideology," "they also function secondarily by repression" (145). On the global scale, these two systems of control, one forcing users to speak and another silencing them, are inseparable, like two sides of a Moebius strip, despite the illusionary division, constitute one surface. It becomes almost impossible to determine where the bigger danger rests. But it is already clear today that everything said—and, even more so, the unsaid—will be held against us tomorrow.

SEXING CYBERWAR

The subject of cyberwar, like the subjects of all wars and like the subject in general, is sexed. This is implicit in all the historical analogies we have invoked; the nomad warriors are men, the *levée en masse* is universal *male* conscription. In this section, we draw on the rapidly growing literature of feminist security studies (Åhäll 2015; Brunner 2013; Enloe 2016; Sjoberg 2014) to consider three hypotheses about the gendering of the cyberwar subject. One emphasizes the exclusion, subordination, and victimization of women in cyberwar. Another, in contradiction, speaks of the emancipatory possibilities such war opens for them. Lacan's views on the matter of sexed subjects significantly differ from these two positions. For him, "sexuation" is not related to biological sexuality and gender; instead, Lacan (1999) describes "masculine" and "feminine" ontologically, as symbolic positions that the subject assumes in relation to the laws of universalization. Thus the Lacanian question is whether the subject can escape the universal law, here that of cyberwar and its totalization. "Masculine" logic is the logic of grouping with the same, driven not by the sense of solidarity but rather by growing insecurity toward others, which immediately mobilizes the need to exclude the uncomfortable encounters. "The feminine" position is not clearly defined: these data subjects are resistant to accept their imposed data identity as a solid, recognizable, readable, stable spatial and temporal data pattern. Instead, they investigate their data representation as, always, a misrepresentation, an inconsistency in data identity. These data subjects, no matter their sex and gender, are not (or are not fully) the subjects of the "universal" laws of cyberwar. From this perspective, we find more promise in a third perspective that takes up the issue of the sexed subjects of cyberwar in the broader context of the relation of information warfare to neoliberal capitalism.

In many ways, cyberwar interpellates women as subjects of exclusion and victimization. It hyphenates two domains, cybernetics (or more broadly computing science) and war, that have both traditionally been culturally coded as male (Hicks 2018). This is so despite the fact that women were present at the origins of cyberwar. Female "computers" programmed the ENIAC, the mainframe machine used in the design of the atomic bomb (Abbate 2003). About eight thousand women worked in Bletchley Park, the site of the British "Enigma" program of computerized cryptoanalysts during World War II, operating cryptographic and communications machinery, translating documents, and performing traffic analysis and clerical duties (Burman 2013). Also during the Second World War, a majority of NSA cryptoanalysis personnel were women (Budiansky 2016; Mundy 2017).

However, this female presence at the origin of cyberwar was subject to a triple erasure. First, the overall masculine path of military–technoscientific development ensured women working on early computer projects were mostly subordinated to, and hence eclipsed by, men. Second, most were at the end of the war rapidly replaced either by men or by computers overseen by men. Third, their contribution was then largely forgotten, in one of the "invisibilizing moves" by which women have repeatedly been "written out of the histories of war" (Sjoberg 2014, 148). Both the military origins of cybernetics and the forgetting of women's part in those origins became part of a series of self-reinforcing cultural feedback loops that constructed computing science as a predominantly and "naturally" male discipline. This has in turn ensured that the central, or at least most prominent, subjects of cyberwar, hackers, are predominantly male.

This gendered construction of computing science would in the United States and Europe from the 1970s onward be challenged by successive waves of feminist technoactivism and advocacy, which, for a time, seemed to make some ground. Donna Haraway's famous call in 1984 for a "cyborg" socialist feminism that would make digitalization "unfaithful" to its military–industrial origins named this this apparently rising trend. It is therefore ironic that from the time of the publication of "Cyborg Manifesto" in the mid-1980s, women's enrollment in U.S. computing programs began a long decline from which it has not recovered. Paradoxically, the cause for this may lie in the expanding popular use of personal computers, which entered households under the auspices of corporate marketing campaigns targeting them to men and boys and linked with an initially highly militarized and masculinized video game culture that gave young men an intense head start in all things digital, independent of formal schooling (Kline, Dyer-Witheford, and de Peuter 2003).

Today computing science, and cybersecurity in particular, remain male professional fortresses.[13] This is the case even while women are, in many parts of the world, highly active in social media and other aspects of networked communication. It is therefore possible to present the gendering of cyberwar as a process in which the digital domains of a femininity acculturated to civilian conversation are turned into battlefields and devastated by high-technology militarized masculinity. This analysis can be supported by pointing out that not only the participants in cyberwar but some of its characteristic practices are gendered. For example, "trolling" is characteristic of misogynist and homophobic digital discourse; there is a manifest overlap between the rise of cyberwar and surging toxic masculinity on Twitter and other social media, sometimes termed "the cyberwar on women" (Thistlethwaite 2016). In the Syrian civil war, female human rights activists using social media in protesting atrocities and documenting the use of rape as a weapon are at risk from government online surveillance and from the hacking of pro-Assad digital militias, such as the Syrian Electronic Army, gathering personal information that puts their lives in danger and "wreaking havoc in online spaces" (Radloff 2012). In this aspect, the cyberwar apparatus's interpellation of its subjects recapitulates in a new technological register the classic binary of male soldiers and female civilians, violent masculinity and pacific femininity, violating men and violated women.

But cyberwar also has a countertendency to hail women as important, even indispensable, protagonists. If social media are a battlefield, trolling is only one of the tactics deployed on it. Others, including viral appeals against enemy atrocities or for defense of the homeland or various forms of "psychological operations," mobilize sentiments and aptitudes associated with traditionally female subject positions attuned to interpersonal and affective interactions. In the Syrian civil war, between

2013 and 2014, a pro-government hacking group operated a scheme by which a female avatar would contact male rebel officers and opposition members on Skype, strike up a conversation, and share a "personal" photo with them; "before sending the photo she typically asked which device the victim was using—an Android phone or a computer—likely in an effort to send appropriately tailored malware" (Reglado, Villeneuve, and Scott-Railton 2015). Once the target downloaded the malware-laden photo, the hacking group accessed his device, "rifled through files and selected and stole data identifying opposition members, their Skype chat logs and contacts, and scores of documents that shed valuable insight into military operations" (Reglado, Villeneuve, and Scott-Railton 2015). Who was behind the "female" avatar is uncertain, but "she" spoke from one of the stereotypical positions assigned to women in war, that of the femme fatale seductress-spy. Other examples are less ambiguous and more innovative. We have already referred to the important role played by female bloggers in Gaza in galvanizing international outrage at Israeli air bombardment. Many of the Israeli military personnel countering their efforts were also women; the IDF's Spokesperson's Unit reportedly had a high proportion of young female soldiers in its ranks, who senior officers considered especially adept at fighting a war for networked public opinion (Patrikarakos 2017).

In her study of "sexing war," Linda Åhäll (2015) addresses the increasing recruitment of women by militaries around the world. She suggests that war waged across networks is one of a number of factors reducing the importance of "brute strength" in soldiering and blurring lines between peace and war. This means that "exclusion policies keeping women out of certain positions within the armed forces have become more and more difficult to justify and . . . more and more positions open up to women" (3). She cites as an example the opening to women of combat roles in the armed forces of the United States and other countries, but the process is also marked in cyberoperations, where it may even be breaching the divide that has segregated women into social media roles and men into programming positions.

One of the very few specific studies of gender in cyberwar, by Alexandria King-Close (2016), argues that U.S. Cyber Command and other branches of the Pentagon have in recent years made strong efforts to recruit women to cyberoperations. About one-third of its employees, and more than one-quarter of its "professional computing" jobs, are held by women, a proportion higher than the overall percentage of women working in information technology roles in the United States (King-Close 2016, 38). In the United States, recruitment of hackers by the state has now become a national security problem, because they can get higher wages in the corporate sector (Slaughter and Weingarten 2016). The Pentagon is taking emergency measures. One is the increased use of privatized contractors (which brings with it its own security problems). Another is recruiting women. In 2015, Theresa Grafenstine, the inspector general of the U.S. House of Representatives, spoke to the issue: "We are absolutely in the middle of a cyberwar," she told a congressional briefing. "It's a new cold war." Presenting data demonstrating that America's cybersecurity workforce is dominated by men, she declared, "If you think we're going to win this war with only half our army—they're going to eat our lunch!" and robustly suggested it was time to "slap Cinderella with a laptop" (Bratton 2015). Thus Haraway's "cyborg" feminism is now exhorted to recover its faith in the military–industrial complex.

In this context, where enlistment as a cyberwarrior is promoted as a feminist path to computational gender equity, the complex work of Elgin Brunner (2013) on the interaction of gender, military institutions, and neoliberalism is especially important. She analyzes the gender assumptions embedded in the discourse and practice of "information war," "perception management," and psychological warfare operations by the U.S. Army in Iraq, operations that now would be considered to fall squarely in the zone of computational propaganda. Her argument is that, regardless of the personnel carrying it out, information warfare's premises are gendered, insofar as they "recur to the most basic stereotypes about what masculinity and femininity imply, namely the holding of and subordination to power," whereby "the Self is hegemonically masculinized and the Other is feminized" (106), with the "capacity to influence . . . among the active and shaping qualities associated with masculine power and skill" and the "quality of being influenced has a negative connotation . . . associated with feminine subordination" (106). In military terms, to be one of "us" is to be automatically masculinized vis-à-vis an enemy who is always, structurally "feminized."

Brunner's (2013) argument is thus that military institutions are so deeply and historically sexed in their conceptions of power that they are "masculinized" regardless of the gender of concrete participants. War is a "gendered identity performance of statehood" (8), a mustering of subjects under the sign of a masculine homogeneity. In this sense, the military is an institution of "hegemonic masculinity"—"structured not only by the hegemony of masculinity over femininity, but also by the domination of one masculinity over other masculinities"—and in that regard it "has not changed since the integration of women" (24). The new emphasis in cyberwarfare on networked organization, flexibility, technological competence, and speed, which in some wars open the fields of "psyops" and "information warfare" to women, only modulates this deeply sedimented structure. This shift Brunner sees as congruent with the rise of neoliberalism, in which the celebration of the network-connected free market hides a subtext of rampant interstate antagonism, such that military information warfare overlaps with economic competition. Her analysis thus joins that of other feminists who point to the very specific shaping and limitations of gender-equity initiatives, and the push for diversity and inclusion, in the context of neoliberalism and militarization (Fraser 2013; Enloe 2016).

Much of the analysis of the gendering of cyberwar addresses its U.S. iterations: there is little or none available in English on the place of women in Russian, Chinese, Iranian, or other cyberwar apparatuses, where their presence may take quite distinctive inflections. Acknowledging this as a void in our understanding of cyberwar subjectification, we now want to take another direction. For if, as Brunner (2013, 104–6) argues, the drive for "technological omnipotence" is a deep dynamic of masculinization, militarization, and neoliberalism, this raises the possibility that the human participants in cyberwar, however gendered, may be overtaken by the intensifying production of automated cyberwar assemblages.

AUTOMATIC SUBJECT

We have argued that cyberwar entails a mobilization of networked subjects, a digital *levée en masse* or militarized "general intellect." This is, however, a paradoxical process because of its relation to automation. Historically, the total war enabled by mass conscription became an ever more mechanized war; in capital, the "general intellect" is activated to automate production, including ultimately its own production, creating a system in which, while human presence remains, it is only as a "link" between machinic components (Marx 1973, 691, 693). Cyberwar, at the cutting edge of a highly automated phase of capital, is fundamentally a process of machinic mobilization, in which humans increasingly play the role of relays within processes whose speed and complexity are deeply inhuman, or at least ahuman. The subjects of cyberwar are thus ultimately not so much humans as they are machine networks to which humans are the most easily compromised point of access. There are several faces to this: the deployment of bots in computational propaganda; the ever-intensifying automation of swarming DDoS attacks; the deployment of massive criminal/military botnets; and the application of advanced forms of artificial intelligence to cyberwar attack and defense.

A chatbot (aka bot, interactive agent, or artificial conversational entity) is a computer program that, through speech or text, converses online. While some bots, such as Apple's Siri or Amazon's Alexa, are overtly machinic, others, often encountered online in Twitter feeds or chat rooms, masquerade as human. Their capabilities range from simply retweeting messages or following a purportedly popular account or mobbing a selected target to convincingly simulating a human interlocutor. They can be preprogrammed with a limited repertoire of conversational gambits, but advanced forms learn from their networked environment. In the infamous case of Tay, a Microsoft chatbot meant to emulate the conversation of a teenage girl, exposure to U.S. internet exchanges resulted in prompt acquisition of misogynist, anti-Semitic, and racist conversational tropes. Some social media companies not only permit bots but tacitly encourage them by making automation easy, as they build traffic volume and create the appearance of vibrant activity, which is good for market valuation.

Twitter has made itself especially bot-friendly; recent estimates suggest that as much as 50 percent of its traffic is automated (Gorwa 2017).

The ubiquity of chatbots means much cyberwar interpellation is performed by machinic agents that can, at least temporarily, pass the Turing test. Bots are generally believed to play a significant role in computational propaganda campaigns emanating from Russia, such as those directed at Estonia in 2007, Georgia in 2008, and Ukraine from 2014 onward. Twitter has identified more than fifty thousand accounts that were engaged in "automated, election-related activity originating out of Russia" during the 2016 U.S. presidential race (Machkovech 2018). These are far from the only examples. In 2017, a political crisis in the Persian Gulf area saw Saudi Arabia and the United Arab Republic accuse Qatar of sympathy with Iranian-supported terrorism and insurgencies. Qatar claimed this was a smear campaign. According to an analysis by *Washington Post* journalist Marc Jones (2017), 20 percent of the active anti-Qatar Twitter accounts were bots, "posting well-produced images condemning Qatar's relations with Hamas, Iran and the Muslim Brotherhood." The bots also "singled out Qatar's media channels as sources of misinformation," tweeted support for the Saudi monarchy, and, during the Riyadh United States–Saudi Arabia summit, "posted thousands of tweets welcoming Trump to Saudi Arabia." Jones (2017) comments that this "mobilization of Twitter bot armies," rather than conveying "an organic outpouring of genuine public anger at Qatar," showed that "an institution or organization with substantial resources has a vested interest in popularizing their criticism of Qatar."

A higher level of cyberwar automation is the DoS attack, in which the perpetrator tries to make a network resource, for example, a website or email service, unavailable by flooding the target with messages. In a DDoS attack, the incoming messages come from many different sources, making the assault harder to counter. DDoS attacks are weaponized machinic interpellation, "hailing" the target so many times that it malfunctions in attempting to reply. From their origin, DDoS attacks have involved software tools for scheduling and sending repeated messages to the target. However, these tools were often used manually, by a single user seated at his own computer, so that successful attacks required large numbers of active human participants. In this form, DDoS could be (and by some still are) seen as a mode of digital insurgency against state and corporate power: examples include DDoS attacks made by the group Electronic Disturbance Theatre from 1998 to 1999 in support of the Zapatistas, using the FloodNet program, and Anonymous's Operation Payback and Operation Avenge Assange a decade later, attacking the sites of corporations attempting to shut down radical organizations, such as the Pirate Bay and WikiLeaks, with its Low Orbital Ion Cannon DDoS tool (Sauter 2014, 109–35; Deseriis 2017).

Such attacks have, however, rapidly entered the repertoire of state power. Britain's Government Communication Headquarters (GCHQ) reportedly retaliated against Anonymous with a series of DoS attacks against its servers in what it called Operation Rolling Thunder (Sauter 2014, 146). DDoS operations are now a common feature of interstate cyberwar, using increasingly automated tools. Evgeny Morozov (2008) narrates how he "signed up" to become a "soldier" in what are widely believed to be Russian-sponsored DDoS attacks on websites in Georgia during the 2008 war between the two nations. This enlistment, far from involving hacker expertise, involved only surveying "the Russian blogosphere," getting directions to a designated website, downloading some easy-to-use software, and, from a pull-down menu, selecting targets— "the Ministry of Transportation or the Supreme Court?"—and clicking "Start Flood." One of Morozov's points is that his experience undermines the assumption that DDoS attacks coming from Russia are necessarily performed by highly trained state operatives; the process is easy for "patriotic hackers" with relatively low digital literacy to undertake, whether spontaneously or through arm's-length state orchestration. In this form, DDoS attacks can be considered a classic instance of the digital *levée en mass*. But what makes this feasible is automation: "war at the touch of a button" (Morozov 2008).

The swarm logic of the DDoS attacks is preserved but raised to a higher level in botnets (from "robot" and "network"), networks of computers (servers, desktops, laptops, smartphones) penetrated by malware so their operations can be controlled by a third party as "zombie armies." Botnets have existed for decades, controlling devices in numbers from the tens and hundreds of thousands to millions. The "botmaster" has at her disposal huge amounts of computing power and bandwidth that can be directed not just to execute DDoS attacks but also to steal data or implant malware. Sometimes participants volunteer to have their computers infected: Anonymous used a type of botnet in the operation of its Low Orbital Ion Cannon, and so did the student group Help Israel Win, which attacked pro-Palestinian websites during the 2009 Gaza War (Shachtman 2009). But users usually do not realize their devices have been compromised. There is a thriving dark web market for purchase or rental of off-the-shelf botnets running on the computers of unknowing participants (Keizer 2010). In such systems, the human is recruited by a phishing attack that will open new networks to viral infection, enlisted only as a vector for access to machinic power.

Botnet capacities are now amplified by the advent of the Internet of Things (IoT), that is, the embedding of networked digital sensors in industrial infrastructures, surveillance cameras, thermostats, baby monitors, televisions, and refrigerators that can communicate with one another, all of which can be used as botnet components. In 2016, the Mirai botnet, drawing on the digital "firepower" of such devices, launched some of the most disruptive DDoS attacks ever known, temporarily shutting down Dyn, a company important in directing internet traffic, and largely knocking the country of Liberia offline. Other "IoT cannons" are almost certainly being produced (Krebs 2016). While to date, botnets such as Mirai are principally used for criminal purposes, the membrane between crime and war is (as we discuss in the next chapter) highly permeable. It is thus almost certain botnets already have a place in military cyberarsenals.

In 2017, the U.S. government's Computer Emergency Readiness Team (2017) issued a warning that North Korean cyberwar plans (given the orientalist code name Hidden Cobra) included a DDoS botnet infrastructure, Deep Charlie, described as "capable of downloading executables, changing its own configuration, updating its own binaries, terminating its own processes, and activating and terminating denial-of-service attacks." In other words, it was a highly automated, semiautonomous system. While such reports can be suspected of recapitulating the notorious "weapons of mass destruction" alarms that preceded a U.S. attack on Iraq, they are not completely implausible. The alert was issued in a context where U.S.

cyberattacks on North Korean intelligence agencies were reported to proceed "by barraging their computer servers with traffic that choked off Internet access" (DeYoung, Nakashima, and Rauhala 2017), so "stack versus stack" (Bratton 2016) hostilities between the two nations may include covert botnet wars.

A striking example of the automated power wielded by nation-state cyberwar apparatuses was the appearance in 2015 of China's Great Cannon, a programming exploit that redirected internet traffic intended for Chinese websites and used it to flood the servers of websites critical of China's internet censorship policies and hosting censorship-evading tools. Not a botnet but a "man in the middle operation," in which the attacker secretly intervenes in communication between two parties, the Cannon hijacked the communications of unknowing bystanders to give the Great Firewall system of censorship and monitoring a massive offensive capacity (Goodin 2015). DDoS attacks, once the weapons of nomadic and anonymous digital rebels, are now, with massively enhanced automation, part of the arsenal with which states quell and punish such revolts. U.S. and British intelligence agencies have developed capacities to redirect internet traffic similar to those of the Great Cannon, not necessarily for DDoS attacks but as a surveillance method, diverting messages to secret servers that impersonate the websites the targets intended to visit (Weaver 2013).

These examples, are, however, dwarfed by the prospects opened by deployment of new forms of artificial intelligence (AI), such as machine learning, for cyberwar. In 2014, Edward Snowden said in an interview that when he quit the NSA, it was working on a cyberdefense system named MonsterMind that would "autonomously neutralize foreign cyberattacks against the US" (Zetter 2014b). It would operate through algorithms capable of rapidly analyzing huge databanks recording internet traffic patterns, differentiating normal flows from anomalous or malicious activity and instantaneously blocking a foreign threat. Snowden expressed anxieties not only over the scale of traffic monitoring required but also because automated responses might extend beyond defense to retaliatory "hackback" strikes against an attacker. Little more has been heard of MonsterMind and how far, if at all, the NSA advanced with the project. Nonetheless, its described modus operandi is consistent with mounting interest by cybersecurity experts in using machine learning and other forms of AI to detect and respond to unusual computer events that might indicate an attack, rather than relying on the static defense of preprogrammed firewalls (Ward 2017; Rosenberg 2017). This interest is itself a response to the perceived capacity AI gives attackers to rehearse exploits and accelerate the speed of intrusions (Yonah 2018).

As early as 2012, a cybersecurity firm study suggested that more than 51 percent of internet traffic was nonhuman; it claimed that of this traffic, more than 30 percent was "malicious," involving activities such as "'spies' collecting competitive intelligence" and "automated hacking tools seeking out vulnerabilities" (van Mensvoort 2012).[14] Other studies push the date machine-to-machine communication predominates further off—but only into the imminent future (Dolcourt 2017). Therefore, doing some violence to Althusser's interpellation, we can say that much of the "hailing" of subjects conducted by the cyberwar apparatus is performed, not just between machines and human subjects—as with chatbots—but between machines and machines. The IoT comprises devices talking to devices. These Things can be made to act as Weapons; the IoT is also a Web of Weapons. There is today widespread discussion of "autonomous weapons" and the tendency to take humans "out of the loop" of automated warfighting systems (Singer 2009; Scharre 2018). While missile-firing drones and armed robots are dramatic examples of this drift, invisible processes of cyberwar may be its forward edge. Capital develops its most advanced technologies in warfare; in cyberwar, we see it advance toward a concrete actualization as what Marx (1977, 255) called an "automatic subject" or what Liu (2010) terms the "Freudian robot."

UNCONSCIOUS WAR

Within cyberwar apparatuses, humans, for the moment, remain a necessary link or relay enlisted in multiple ways, voluntary and involuntary. Yet while humans remain in the loop, or on the loop (that is to say, with a veto on otherwise automatic processes), it is within a war-fighting system that increasingly decenters subjectivity as a "peripheral" (Gibson 2015). Because of this, the human subject of cyberwar is dazed and confused.

This is in part a consequence of the intentional secrecy of cyberwar, but the possibilities of such stealth, and its intensification by contingency and accident, arise from the speed, scope, and complexity of the technology of cyberwar apparatuses.

Deeply implicated as users are in the militarization of networks, their involvement is frequently unknowing or misrecognized. We are indeed "empowered" by technology—but not necessarily in the way we are told. Rather than acting as globally aware networked individuals, intervening purposefully in great political events with a few deft touches to an iPhone, our cyberwar involvement is as likely to be a misapprehending, deceived, or involuntary conduit for war whose outbreak has either passed by unnoticed or was only imagined (at least until this imagined onset provoked real counteraction), or whose combatants are drastically misidentified. In conflicts where a crucial action may be the opening of virally contaminated email, the retweeting of a message from a software agent mistaken for a human, or the invisible contribution of a hijacked computer (or digitalized refrigerator) to a massive botnet, we are in the realm of Marx's "they do it, but they do not know it." "Even if you do not see the war, the war sees you" is the logic of the blind gaze of cyberwar, a regime in which although "the subject does not see where [this regime] is leading, he follows" (Lacan 1998, 75).

The obscurity inherent to cyberwar afflicts even those most expert in its prosecution. During the U.S. occupation of Iraq, the CIA and Saudi Arabia's intelligence service set up a "fake" jihadi website to monitor Islamic extremist activity. In 2008, the U.S. Army and the NSA concluded that the "fake" site was actually serving as an operational planning hub for attacks by Saudi Arabian jihadists joining the Sunni insurgency. When they proposed the site be destroyed, the CIA objected, but Pentagon hackers proceeded with the "take-down." They inadvertently disrupted more than three hundred servers in Saudi Arabia, Germany, and Texas. As a task force participant ruefully explained, "to take down a Web site that is up in Country X, because the cyber-world knows no boundaries, you may end up taking out a server that is located in Country Y." The Saudi Arabian intelligence service, which regarded the "fake" site as a "boon," was furious; mollification required "a lot of bowing and scraping." The

CIA, too, was resentful; the agency "understood that intelligence would be lost, and it was; that relationships with cooperating intelligence services would be damaged, and they were; and that the terrorists would migrate to other sites, and they did" (Nakashima 2010).

A more serious example of unintended consequences is Stuxnet, the computer worm planted in the computers at the uranium enrichment plant outside Natanz to prevent Iran from building a nuclear bomb, an operation now widely attributed to a joint U.S.–Israeli intelligence operation. As we noted in chapter 1, the worm's impeccable simulation of a mechanical failure apparently unrelated to software performance is considered a watershed in the development of cyberweaponry. What it is not so generally recognized, however, is that it went out of control. Stuxnet's discovery by the security company VirusBlokAda in mid-June 2010 was the result of the virus accidentally spreading beyond its intended target due to a programming error introduced in an update. This allowed the worm to enter into an engineer's computer connected to the centrifuges and thence travel to the internet. It then propagated to industrial sites far from Natanz, not only in Iran but in Indonesia and India, and beyond, reportedly infecting the systems of oil giant Chevron and a Russian nuclear plant. As one cybersecurity expert puts it, "By allowing Stuxnet to spread globally, its authors committed collateral damage worldwide" (Schneier 2010). Although in many of these cases, the virus did not activate, because of differences between the Natanz system it targeted and the others it accidentally infected, another consequence was that the Stuxnet code became widely available for use or adaptation by hackers other than those who developed it. Such probably inadvertent propagation can be considered what Paul Virilio (2000) terms an "integral accident," a malfunction intrinsic to, and inevitable for, viral cyberweapons.[15]

Once one passes to the civilian perception of real or imagined cyberwar effects, the scope for misrecognition increases and potentially ranges from imagining wars where none exists to not noticing those that are actually raging. Zetter (2016b) reports a "misrecognized" attack on a power grid in Ukraine that occurred on December 23, 2015, when twenty-seven substations of the Prykarpattya Oblenergo, a Ukrainian power distributor that serves 538,000 customers, went dead after the company's computers were infected by a version of a high-powered web-based malware BlackEnergy 3, in what is generally regarded as an act of Russian aggression, although the attribution, as always, is inconclusive. The cyberevent attracted the attention of cybersecurity and hacking communities: the blogosphere and specialized online channels and platforms competed for the most informed interpretation of the blackout. In Ukraine, however, where the cyberattack took place, it was unnoticed, despite successfully plunging hundreds of cities and villages into darkness. With the exception of security, administration, and technical personnel of the power station, the local population took the blackout for a common power shutdown, a nationally centralized procedure aimed at saving electricity in the country's declining and war-afflicted economy.

In a reverse example, in August 2008, cyberattacks took place in the midst of a broader armed conflict between Russia and Georgia over the disputed territory of South Ossetia. Although these attacks, allegedly coordinated or encouraged by the Russian state, did not significantly affect the ongoing kinetic action, distribution of malicious software; defacement of political, governmental, and financial websites; and multiple DoS and DDoS attacks on governmental, financial, news, and media websites generated confusion and panic among the population of the country at a time when "Georgia was the most dependent on the availability of information channels" (Tikk, Kaska, and Vihul 2010, 69–79, 72). Then, on March 28, 2011, the internet in Georgia and Armenia went down for nearly the entire day after a seventy-five-year-old Georgian woman named Hayastan Shakarian, while digging for scrap copper, accidentally cut a fiber-optic cable owned by Georgian Railway Telecom that runs through the two countries (Millar 2011). It would not have been too strange if, to a traumatized wartime population, this accident had signaled another kinetic offensive (Deibert 2013, 29). How many times would such suspicions need to be shared and commented on in social networks to become someone's "knowledge"? To scale and speed up to the status of "fake news"? To serve as a useful context or leverage for a future cyberattack? To premediate an invasion?

The cybernetic autopoiesis of unplanned and undesired incidents, unavoidable and unpreventable accidents, as well as the masterminded and preplanned operations constitute the ongoing production of events and semblances constitutive of cyberwar dynamics. Everything, even what did not have place, did not happen, or was misattributed, has a positive value in the cyberwar economy. This trompe l'oeil creates blind spots in the field of vision of all observers of cyberwar.[16] It accelerates what Žižek (1999, 322) calls the "decline of symbolic efficiency" in digital capitalism. As Jodi Dean (2014, 213) explains, this develops the Lacanian idea that

> there is no longer a Master-Signifier that stabilizes meaning, that knits together the chain of signifiers and hinders their tendencies to float off into indeterminacy. While the absence of such a master might seem to produce a situation of complete openness and freedom—no authority is telling the subject what to do, what to desire, how to structure its choices—Žižek argues that in fact the result is unbearable, suffocating closure.

A "setting of electronically mediated subjectivity [that] is one of infinite doubt and ultimate reflexification" intensifies "the fundamental uncertainty accompanying the impossibility of totalization" in a symbolic environment where "there is always another option, link, opinion, nuance or contingency that we haven't taken into account" (Dean 2014, 212). Computational propaganda that aims to mystify invasions and occupations, or promote cynical disaffection from an adversary's political system, actively weaponizes the "decline in symbolic efficiency," but it is endemic to the whole field of cyberwar.

The extreme uncertainty and opacity of cyberwar do not, however, inhibit the interpellative effects of contending cyberwar apparatuses as they summon up cybersoldiers, patriotic hackers, vigilante militias, and security-conscious digital citizens. On the contrary, the problems of verifying or disproving multiple alarms and accusations accelerates these processes and puts them into overdrive. To put this point in psychoanalytic terms, as we noted previously, commentators on Althusser have criticized the appropriation of Lacan's theories of the subject in his account of ISAs. These critics point out that what Althusser misses in Lacan's account is that the subject is *always incomplete*; it is precisely what can never be fixed by a specific subject position or identity. However, the implication of this incompletion is not that the subject remains some untouched and primordial haven of authenticity but rather that this lack drives to ever more compulsive (because unfulfillable) attempts to attain a definitive identity. Translating this into political terms, we would say that it is the inescapably incomplete, provisional, and easily falsified nature of all accounts of cyberwar that energizes the adoption of increasingly militarized, extreme, paranoid, and unshakable subject positions vis-à-vis its alleged events.

For example, shortly after the outbreak of the rebellion that grew into the Syrian civil war, there was an abrupt but near-total shutdown of the Syrian internet. A common assumption, at least in the West, was that this was an attempt by the Assad regime to black out online dissent, as Mubarak had attempted in Egypt. But according to Edward Snowden, the event was caused by intrusion into the system conducted by the NSA—not intentionally, however, but by accident, in a botched hack of the Syrian state's communication and electronic defense system (Ackerman 2014). Whereas the first attribution cast the Assad regime in the conventional role of despotic suppressor of civil rights, rightly opposed by liberal democracies, the second reversed the significance of the blackout, making it evidence of—once again—NSA cyberaggression against foreign states, and incompetent aggression at that. But those opposed to this characterization could point out that at the time Snowden made his diagnosis, he was reliant on Russia, a supporter of the Assad regime, for political asylum. The blackout of Syria's internet connection thus also becomes an epistemological blackout about its cause, a blackout in which every initial position on the politics of Syria's civil war could be preserved and reinforced.

To provide a final example that is closer to home for many readers, as we suggested in chapter 1, there is now fairly convincing evidence that Russian intelligence agencies, whether directly or by proxy, attempted some intervention in the 2016 U.S. presidential election by way of "fake news." It is also clear that some of the news reports claiming to substantiate or expand this claim, by claiming, for example, to detect Russian hackers in Vermont's power grid or by broadly characterizing a sweepingly wide range of U.S. media outlets as accomplices of Russian cyberwar, are inaccurate and tendentious. The abyss of this double falsification—"fake news" compounding "fake news"—becomes a zero-gravity free-fire zone within which contending factions within the U.S. political system trade charges of treason, producing a civil war effect possibly beyond the wildest dreams of the toilers at the dreary offices of St. Petersburg's Internet Research Group.

SPEAKING OF DREAMS

Despite the warnings, protocols, or simply general situational awareness regarding usage of social media and mobile phones in the military zones, soldiers seem to be no different from other users, for whom sharing activities with mobile apps has become the way of living socially and communicating, disclosing one's everydayness to the network's gaze. Today, when the topic of security is trending and exploited, users can not be fully unaware of the scope of their disclosures, the associated risks and potential harm. Perhaps this is precisely what often drives the ongoing ubiquitous disclosure: exposure hurts and pleases users at the same time—the contradictory sensation that Lacan named by the French term *jouissance.*

In Lacan's terminology, the structure that hosts the alienated subject embodies "the law of the Father."[17] However, as Kittler (1997, 140) pointed out, "the world of the Symbolic [is] the world of the machine." We, too, consider that the Lacanian term can be applied to the protocological and algorithmic logic and logistics, the law and order of network operations, to which the user must submit, even if by means of pretense, in order to become *a user* in the first place. Lacan saw the subject's inability to resolve the contradictions of this dual position as nominally free, but actually unfree, as leading to the forced choice between two options: either becoming a "dupe" who enacts a misunderstanding of the system, while having some understanding of it, or a "nondupe," the one who believes in controlling the system by reducing computer to source code (think of a hacker, a corporate CEO, or the internet libertarian, fetishizing code as a solution to anything) (Chun 2008, 300).

As cyberwars intensify, digital networks become harder and more hazardous to use due to the continuous blocks, shutdowns, intrusive surveillance, new toxic malware, and so on. And yet, the internet in the mind of the majority of users remains conceived via the idea of "connection" rather than, say, "antagonism" or "collapse." The user thus enacts the imaginary relation with technology promised by commercials (the real "fake news" in consumer society). As such, the user, who is not fully unaware of the problems, yet dismisses them, is the necessary link in the cyberwar assemblage. The user is forced to act *stupidly*. This stupidity "is not always at odds with intelligence but can operate a purposeful exchange with its traits. . . . Intelligence itself depends on a withholding pattern that in some cases matches the irremediable reluctance of the stupid" (Ronell 2002, 10).

In regard to the second option of being a "nondupe," Lacan took the notion of the law and order further by tweaking it again to *les non-dupes errant,* meaning "those who do not let themselves be caught in the symbolic deception/fiction and continue to believe their eyes are the ones who err most" (Žižek, n.d.). This perfectly captures the epistemological condition in cyberwar when we have too many obvious proofs, too much information. One is tricked precisely at the moments of clarity or embracing the power granted by the network as well as at the moments of living the most "authentic" yet extremely ambiguous "carnal resonance" (Paasonen 2011) of arousal or the "real affect" of anxiety (Lacan 2016).

On one hand, the user-subject is the subject of data. To rephrase Lacan's definition of the subject (that which is represented by one signifier for another signifier[18]) in the context of information economy, the subject of data becomes that which one data point represents for another data point. Here the user is trapped in the representational data chain as negativity, as a figure of exclusion, whose place is persistently taken over by data: the subject is not present but always already represented. On the other hand, the user-subject is a "suture" that works to bridge the gaps, often by imagining connections or relations in operations of the porous "accidental megastructure" of the stack (Bratton 2016), where there are none. Here the subject appears as a figure of recursion, caught by circuits of misrecognition of patterns, as reflections or traits of imaginary identification, "rearranged" or "retranscribed" by the stochastic structure (Lacan 1997, 181) of language or that of the net.

These technological structures provide the subject with "reference points" for identification and orientation so that the subject allows herself or himself "to be fooled by these signs to have a chance of getting [one's] bearings amidst them"; the subject "must place and maintain [oneself] in the wake of a discourse and submit to its logic—in a word, [the subject] must be its dupe" (Miller 1990, xxvii). In a certain sense, of course, it is better to be a dupe of paranoia than a dupe of the technological, linguistic, or ideological system. It is precisely in recognizing one's subjective position as a dupe of the system that the subject secures a possibility of "another knowledge" upon which to act, when the system tightens its grip: "just because you're paranoid doesn't mean they aren't after you." Instead of choosing between a possibility of being paranoid or a possibility of being followed, one refuses to take these as mutually exclusive and to be divided by such choice, especially when the right option to choose *seems* apparent.

In the documentary *Lo and Behold, Reveries of the Connected World* (2016), German filmmaker Werner Herzog converses about the impact of the IoT, AI, and autonomous machines with several distinguished computer and mechanical engineers, scientists, philosophers, and entrepreneurs, including Leonard Kleinrock, Bob Kahn, Elon Musk, Danny Hillis, Sebastian Thrun, and Ted Nelson, whom he invites to share their visions of the future. And then, perhaps with a hidden intention to punctuate or disrupt his guests' technoenthusiasm, Herzog suddenly asks them an awkward question, upon which they inevitably stumble: "Prussian war theoretician Clausewitz," Herzog proclaims, "famously said, 'Sometimes war dreams of itself.' Could it be the internet starts to dream of itself?" Herzog does not explain how this reflection about war that allegedly belongs to the nineteenth-century Prussian general appears in the same sentence with "the internet," but it seems to touch what is already on everybody's mind. So, one response to Herzog's "von Clausewitz question" could be, "Yes, the internet dreams of itself—and when it does, it dreams of war." Such a dream would, however, have to be understood as akin to one on which Freud reported: a father falls asleep near the coffin of his dead child and sees his son alive, whispering, "Father, don't you see I'm burning?" Waking, the father notices the boy's dead body caught on fire from a candle (Freud 1900, 509). To Freud, the dream manifested the father's wish fulfillment, allowing him to prolong his sleep as a means of seeing his child alive. Lacan, however, argued that in dreams, where repressive mechanisms are disabled, we find ourselves in a dangerously close proximity to the unbearable real and wake up to avoid the encounter. To him, our imaginary construction of reality was the waking daydream in which we escape the real. That daydream would be our continuing reverie about the plenitude and peace of the promised digital future, while the unbearable real is war and cyberwar, burning with ever-increasing intensity.

③ *What Is to Be Done?*

ENDOCOLONIZATION AND RECOMPOSITION

Emptied squares, still vigilantly patrolled by police, had been hosed down, the tents and banners gone to the dumpster, graffiti sandblasted, protestors dispersed, subpoenas for social media records sent and security videos scanned; prosecutions were under way. In an article written as Occupy Wall Street and other postcrash protests waned, Benjamin Noys (2013) raises the question about the missing military dimension of contemporary revolt against capital. His starting point is Paul Virilio's reinterpretation of the Marxist narrative of industrial working-class struggle as an encounter between a militarized ruling class, controlling the army and security apparatus, and a proletarian "counterwar machine" prepared to forcibly contest exploitation by sabotage, street demonstrations, and strikes: "In this model the forms of the traditional workers' movement—notably parties and unions—become alternative 'armies' to counter . . . military domination" (Noys 2013). However, according to Virilio, in the latter half of the twentieth century, this form of resistance is defeated by the increasing war power of capital. Its victory is most dramatically signified by nuclear weapons, an apex of destructive power independent of human mass mobilization: "the proletariat's determining role in history stopped with the bombing of Hiroshima," Virilio (1990, 29) declares. But the ascendancy of capital's rule is more broadly represented by its development of a high-technology, and specifically informational, military apparatus, which diffuses into techniques of policing and management, giving global

117

and accelerated powers of command, eroding the possibility of localized resistance in factory or neighborhood.

Virilio terms this *endocolonization,* a process in which the military powers developed in the course of foreign and particularly colonial wars are applied to discipline and reduce the proletariat, so that "one now colonizes only one's own population. One underdevelops one's own economy" (Virilio and Lotringer 1983, 95). As Udo Krautwurst (2007) explains, Virilio throughout his work develops this concept in rather diverse directions, but it broadly "refers to the intensification and extensification of war within and throughout actually existing state forms, an inwardly directed expansion of the principle of the State, manifested in an increasing militarisation of the social." It has two main aspects: "one macrosocial, wherein a war economy is carried over into peacetime, restraining potential development in civil society; the other microsocial, such that the human body is increasingly becoming a site of technology itself" (139)—a digital technology that, for Virilio, has an inescapable military valence.

While Noys (2013) does not necessarily fully endorse Virilio's analysis of high-technology militarism, what interests him about it is how, "in a rather uncanny way," it "dovetails" with the analysis of class conflict presented by Marxist theorizations of the decomposition of the traditional working class from the 1970s on.[1] "Endocolonization" corresponds with the neoliberal discarding of any social contract between capital and labor, the smashing or "hollowing out" of trade unions and labor parties, and a "decoupling [of] the worker from work" that "dispense[s] with the affirmation of worker's identity as an essential 'moment' for capitalist reproduction." "In this situation," Noys notes exegetically, "traditional forms of popular resistance . . . become put into question," apparently leaving only "suicidal" options, such as terrorism. It is this dilemma that marks the limit of the Occupy movements of the 2011 cycle of struggles that, at least in their North American and European versions, either fizzled out in symbolic gestures or resulted in reinstallations of oligarchic capital. The question Noys poses, though does not answer, is whether Virilio's conclusion about the military exhaustion of "proletarian" struggles is correct, or if the "disintegration of old and compromised forms of labour identity" in any way opens the way to new processes of revolution and the invention of a new "counterwar machine."

Cyberwar is a paramount example of "endocolonization," with both "macrosocial" implications, in terms of the construction of new "military–internet" complexes (Harris 2014), at the expense of other social purposes, and the "microsocial" implications of the invasion of the network user by algorithmic surveillance, warbots, malware, and automated propaganda and misinformation. In this chapter, we consider Noys's questions about the decompositionary and recompositionary possibilities of cyberwar-time for anticapitalist movements. Our argument is that the destructive, endocolonizing effects of cyberwar can only be counteracted by explicitly thematizing them as foci of social struggle, making the demilitarization of digital networks an issue around which new oppositional formations can take shape and older ones connect in new ways. We advance this argument under a series of (in)appropriately military headings. The first two sections address the "tactical" responses of contemporary social movements to the new digital powers of militarized capital, in terms of both defensive measures, such as cybersecurity and antisurveillance protection, and offensive capacities, notably the recent surge in politicized hacking. We then move to an "operational" overview of a series of issues that together might constitute a broader campaign against the endocolonizing effects of cyberwar. This is followed by a digression on the "organizational" lessons that may be learned from, and applied to, the cyberwar context, particularly in regard to debates about verticalism and horizontalism, or vanguard and network. Finally, we turn to the "strategic" question of the meaning and prospects of revolution in the context of cyberwars and what it might mean to understand and construct, in the full ambivalence of the phrase, a "counterwar machine."

TACTICS 1: ANTISURVEILLANCE

Left attitudes over digital technologies in meetings, art, agitation, and protests have always shuttled across a Luddite–cyborg spectrum. Suspicion of computers as military–industrial tools, widespread in the 1960s and 1970s, was succeeded in the 1990s by critical net culture (represented by organs such as the tactical media collective Critical Art Ensemble, the open access Public Netbase, Geert Lovink and Pit Schultz's internet mailing list net-time, the German internet magazine *Telepolis,* and the Slovenian media lab Ljudmila) and the alter-globalist "cyberleft" (Wolfson 2014; Apprich 2017). These currents then ebbed, amid growing despondency about the recuperative powers of "communicative capitalism" (Dean 2005, 2009), only to be followed after the crash of 2008 by Facebook revolutions "kicking off everywhere" (Mason 2012). As that cycle of struggles declined, a more somber assessment of the digital returned again. The major cause was state and corporate surveillance.

Recognition dawned for different movements at different moments. China's dissidents had for years been entrapped by, and evading, their nation's Great Firewall of censorship and surveillance. In Egypt, the police killing of radical blogger Khaled Saeed was one of the sparks that ignited the Tahrir Square uprising of 2011. The arrest, torture, and death of protestors identified on social media or mobile phone would become a feature of the Syrian civil war and other aftermaths of the Arab Spring. In Ukraine in 2014, protestors close to the scene of violent clashes in Kyiv received a text message from the security forces who tracked their mobile phones: "Dear subscriber, you are registered as a participant in a mass riot" (Walker and Grytsenko 2014). In the United States, New York City prosecutors' 2012 subpoena of the tweets of an Occupy Wall Street protestor arrested during a mass march on the Brooklyn Bridge was a wakeup call. But it was Edward Snowden's 2013 "no place to hide" revelations about the scope of the NSA's foreign and domestic surveillance programs, and those of allied agencies such as the British Government Communications Headquarters (GCHQ), that in North America and Europe terminally punctured liberal rhetoric about "internet freedoms" and decisively confirmed protestors' mounting apprehensions over the exposure of networked dissent to digital monitoring (Greenwald 2013).[2] Although the USA Freedom Act of 2015 is sometimes trumpeted as having ended the era of mass surveillance, it imposed significant restrictions only on the NSA bulk-data phone dragnet, did little to touch its internet trawling capacities, and left many loopholes; powers for large-scale warrantless surveillance have been extended by the Trump administration (Ackerman 2015; Knibbs 2015; Lecher 2018: Emmons 2018).

Surveillance has chilled internet use in the United States, making the general public more reluctant to visit websites, use search terms, express opinions, or hold library resources on "politically sensitive" topics that might invite scrutiny (Greenwald 2016). For those involved in movements of protest, anxiety is even more justified. How many suspects, involved in what types of activity, are singled out from the wholesale information trawling by the NSA or other high-level national security agencies for specific attention is only fitfully illuminated by courtroom events or freedom of information requests. What evidence is available shows that security agencies' digital watchlists have extended from hacker groups, such as Anonymous, to civil liberties and humanitarian organizations, prominent academics and journalists, and visitors to WikiLeaks (Dencik, Hintz, and Cable 2016, 3). Though this information is partial and episodic, uncertainty about the actual scope of surveillance need not diminish the efficiency of panoptic social discipline; indeed, it is intrinsic to its operation.

A yet blunter threat to social protest is posed by the devolution of surveillance technologies and techniques from elite security forces to increasingly "militarized" police forces (Wood 2014). In the homelands of advanced capitalism, these forces now track mobile phones, collect troves of video, analyze social media postings, and compile profiles for so-called predictive policing. Core police capacities are supplemented by private firms selling off-the-shelf surveillance equipment and social media data to authorities. This apparatus, a translation of cyberwar powers to cyberpolicing practice, is regularly deployed at protests and actions of all sorts and extends its range both to the preemption of such events and to the subsequent tracking of participants (Wood 2014). Western software companies vend profiling and monitoring systems to both intelligence and policing services to customers worldwide, including the most violently repressive regimes on the planet (Deibert 2013).

Because of these developments, cybersecurity, long a concern of "cypherpunk" hackers (Assange 2012; Greenberg 2012), has generalized across movements. In North America, digital self-defense manuals and information sheets proliferate. They come from a variety of sources: veteran digital libertarian movements, such as Electronic Frontier Foundation; anarchist and autonomist hacker groups like Riseup.net (a secure communication collective that identifies its members only by bird names); NGO-oriented organizations addressing social justice activists in the Global

South where "computer operating systems are frequently out of date, software is often pirated and viruses run rampant" (Tactical Technology and Frontline Defenders 2009; Citizen Lab 2017); and recently in the United States, feminist and antiracist organizations mobilizing against the rise of the alt-right and the Trump presidency (Blue 2017; Equality Labs 2017). Through such guides, and in movement-organized seminars and briefings, activists learn methods to reduce their "surveillance footprint." Topics include email encryption; establishing multiple online identities; installation of open source antimalware programs; selecting, maintaining, and remembering secure passwords; recovery from information loss, whether by theft, hacking, or device confiscation; hiding many people under one name; producing online behavior patterns to trick a trained observer; and, equally important, means to effectively and permanently destroy information when necessary.

Various levels of anonymizing technologies from virtual private networks to Tor to Tails, aka the Amnesic Incognito Live System, are discussed. Advice on mobile phone security is usually a priority, from basic precautions about default settings and location functions through comparison of the security features of competing models to the use of burner phones or shuffling SIM cards and "jailbreaking" or "rooting" smartphones to access locked-down configurations. Other topics include countermeasures against trolling, harassment, and "doxing" by online opponents; discussion of various forms of internet censorship and circumvention practices to bypass or overcome it; and the use of special platforms designed to protect whistleblowers. In some contexts, countersurveillance, or *sous-veillance,* also gets attention, particularly in regard to mobile phone video evidence against police violence or spurious charges: "learn how to use live video in situations both ordinary and extreme" (Blue 2017). Advice on prepping for demonstrations may extend to the use of body cameras (including head mounts); necklace cameras (always on); flash drives containing legal briefings; and how to upload videos of crucial events to social media, ensure wide circulation, and track responses.

Such briefings share a common premise: digital communication is increasingly perilous, but too important to abandon, and can be secured by skillful tactics. Tactical Technology and Frontline Defenders (2009), addressing social activists in the Global South, writes,

The goal of this guide is to help you reduce even the threats that do not occur to you, while avoiding the extreme position, favored by some, that you should not send anything over the Internet that you are not willing to make public.

From the very different ambience of San Francisco, feminist digital activist Violet Blue (2017, "Resist") promises that while anti-Trump protestors should start to "think like a hacker," they don't need to "quit social media" or become a "shadowy underground" figure:

> Welcome to modern tech: It's shoddy, it never works right (or it crashes and may blow up), it sells us out to anyone who asks . . . And yet it's the sword with which we've become a credible threat to corrupt politicians and civil rights attacks like state-sponsored censorship.

Antisurveillance sentiment has in some U.S. social movements reached levels where digital self-defense advocates feel obliged to warn against "activists [who] one-up each other about who knows more, or has better 'OPSEC' [operational security] and against 'Charlatans and bad activist advice' such as Kickstarter promises of 'military grade encryption'" (Blue 2017, "Charlatans and Bad Activist Advice"). The latest purported frontier in technical protection is biometric-baffling face paint and "use to confuse" clothing and face paint (Kopfstein 2017).

In their "a user's guide for privacy and protest," Finn Brunton and Helen Nissenbaum (2015, "Introduction") summarize these and similar tactics as practices of "obfuscation" or "the deliberate addition of ambiguous, confusing, or misleading information to interfere with surveillance and data collection" in the time when we are "unable to refuse or deny observation." They distinguish between several kinds of obfuscation practices: selective and general, short-time and long-time, and also those conducted in secret and others conducted openly: "For some goals, for instance, obfuscation may not succeed if the adversary knows that it is being employed; for other goals—such as collective protest or interference with probable cause and production of plausible deniability—it is better if the adversary knows that the data have been poisoned" ("Introduction"). However, as Brunton and Nissenbaum acknowledge, the speed at which these practices become disempowered, often by way of corporate and state power appropriating revolutionary tactics as soon as they emerge, is astonishing. In chapter 2, we identified this process as the consequence of the troubling reciprocity between surveillance and mobilization, one of the typical features of the new type of cyberwars. For example, Twitter, a passive surveillance platform for the Russian government during the Snow or White-Ribbon Revolution in December 2011 (the protests against Russian legislative election results), was fully weaponized within a month. By January 2012, the Twittersphere was infiltrated by pro-government trolls countering critique of the fraud election by massively disseminated praise for and sympathy with Putin; "God save Putin" or "pointless protest" posts scored just as high as "you are (the) movement," "Putin thief," or "be one white-ribbon" (Spaiser et al. 2017, 139, 142), and "the use of Twitter bots" became "a reliable technique for stifling Twitter discussions" (Brunton and Nissenbaum 2015, "Core Cases").

How far the advice of digital self-defense programs is followed is, moreover, dubious. As such programs usually acknowledge, assessing the actual level of security germane to different types of activism, treading a boundary between carelessness and paranoid grandiosity, is challenging and highly context dependent. Advanced security measures require careful attention to use properly, can be compromised by easy errors, are subject to technical failure, and can in and of themselves attract, rather than deflect, attention. A recent report on the actual practices of social movement participants in the United Kingdom is sobering (Dencik, Hintz, and Cable 2016). Based on interviews with activists, it finds that while there is, post-Snowden, a heightened awareness of surveillance, very few act on it. Many organizations are so deeply reliant on networks and social media for "general awareness-raising, advocacy, mobilizing, organizing and expanding their actions and membership" that introducing new technological practices, always in the context of stretched resources, is extremely inconvenient and time consuming. Even when antisurveillance measures do not require a level of expertise exclusive to "techies," users inherit an expectation of seamless, effortless computing, particularly in volatile situations requiring fast communication. Activists may in crisis situations abandon platforms known to be insecure and turn to others, as they did in Egypt when protestors left Facebook in favor of Signal (Mackey 2016). However, according to Dencik, Hintz, and Cable (2016), in day-to-day life, the surveillance issue is most often resolved by "self-regulating online behavior," such as "not saying anything 'too controversial' on social media" or by assertion that an organization has "nothing to hide." This latter assurance may be either overconfident or an indication of panoptic political chill. It also ignores the fact that ubiquitous surveillance has gone far beyond users' verbal expression and use of keywords to the recordings of location, chronology, and activity now embedded in our machines; for example, those who attend a protest or are close to a terrorist attack will always be the subjects of data that positions them in proximity to such events.

Though Snowden's revelations have had an effect on popular perception of digital networks, their implications are easy to misrecognize. In North America, they have been cynically played by the giants of digital industry. After years of voluntarily complying with the NSA PRISM surveillance program, providers of digital platforms and services responded proactively to the new antisurveillance sentiment by offering updated services with reinforced security features. This swerve has led digital capital into some conflicts with state security, most notably in the legal dispute between Apple and the FBI arising from the company's refusal to construct a "backdoor" to undo the password on an iPhone5c used in a 2015 terrorist attack (Zetter 2016a). Privacy activists supported Apple, but before the case could go to court, the FBI paid to crack the password with the help of an outside contractor (variously reported as either a professional hacker or an Israeli cybersecurity firm), thereby evading any legal showdown between the intelligence agencies and their erstwhile corporate partners. Such contradictions between free enterprise and the security apparatus that protects it may create certain technological options for movement activists—at least those who can afford the latest upgrades. However, information capital's new security consciousness hijacks post-Snowden discourse to evade its larger implications. It is intended to deter mass defection from mobile phone and social media, keep the public on the networks, and hence within a field where big data about habits, purchases, locations, friends, and contacts continue to be amassed by corporations whose privacy assurances can and will be compromised by voluntary or forced collaboration with state authorities—or ingeniously bypassed.

There are, therefore, serious limitations to purely tactical, and technical, movement adaptations to the surveillance state. An alternative is political and legal contestation of surveillance. In North America, some of the most important of these challenges come from Afro-Americans, First Nations, and Arab-Muslims, who, on the basis of collective historical experience, point out how profiling systems at once constitute and control specific suspect racialized social groups (Kundnani and Kumar 2015), demonstrating how Virilio's "endocolonization" is most virulently applied to the already colonized. In many cases, such challenges also point directly to the military aspects of racialized surveillance. Thus the digital tracking of Black Lives Matter activists and leaders by both state agencies (Joseph 2015) and private cybersecurity firms (Buncombe 2015) stands as an extension of the paramilitary combination of "stingray" mobile phone signaling interceptions, drone observation, and information "fusion center" cross-checking transferred from Middle Eastern wars to the policing of Afro-American neighborhoods (Collins et al. 2015), practices that in turn build on a lineage of white vigilance rooted in fear of slave revolt (Browne 2015). At the time of writing, two activist groups, the Color of Change and the Center for Constitutional Rights, are suing the FBI and the Department of Homeland Security on the surveillance of protests in eleven cities, arguing that it undermines free speech while serving to "chill valuable public debate" (Timberg 2017).

In Dakota, the company TigerSwan, founded by a former U.S. special operations soldier, was hired by Energy Transfer Partners to monitor the Sioux Nation's Standing Rock anti-oil pipeline protests. It conducted radio, video aerial, and social media surveillance of protestors; shared intelligence with police; and mounted online campaigns to discredit the movement. TigerSwan now faces civil suits for operating a private security service without a license (Brown 2017; Brown, Parrish, and Speri 2017). In another case, a coalition of activists, including high-tech workers, has protested the potential involvement of Peter Thiel's software company Palantir in building a "Muslim database" to support Trump's promised "extreme vetting" of immigrants (Woodman 2016; Buhr 2017). Palantir as a start-up benefited from investment by CIA venture capital front company In-Q-Tel and has a record of work with the CIA, NSA, and U.S. Customs and Border Protection Agency (Biddle 2017).

Over summer 2018, a remarkable wave of technology-worker resistance to military and militarized policing projects swept through Silicon Valley (Tarnoff 2018). At Google, workers successfully organized to shut down Project Maven, a Pentagon project that uses machine learning to improve targeting for drone strikes. Following this, Google's CEO, Sundar Pichai (2018), published a statement of principles on AI development, saying it would not work on AI weapon or surveillance contracts, although reserving the right to pursue cybersecurity projects. In June, the company withdrew its bid on a $10 billion Pentagon contract for the Joint Enterprise Defense Infrastructure (Jedi) cloud computing project (BBC 2018f). Meanwhile, at Amazon, workers petitioned Jeff Bezos to stop selling the corporation's Rekognition facial identification software to U.S. police departments and the Immigration and Customs Enforcement (ICE) agency, notorious for its "zero-tolerance" enforcement policies (Conger 2018a). At Microsoft, workers similarly demanded the termination of a $19.4 million cloud computing contract with ICE (Frenkel 2018), and at Salesforce, workers tried to block the company's involvement with Customs and Border Protection (CBP).

Other legal challenges to military-derived surveillance have come from individuals, such as the case launched by David Carroll, in U.K. courts, against Cambridge Analytica, a former military contractor that moved into the business of voter profiling. Analytica is widely believed to have assisted the Trump campaign in targeting ads to select voter groups during the 2016 election. Carroll's case attempts to use British privacy law to compel Analytica to reveal the basis on which his (and, by implication, other voters') personal profile was constructed (Cadwalladr 2017a). Such examples move beyond digital self-defense toward what Dencik, Hintz, and Cable (2016) call "a (re)conceptualization of resistance to surveillance on terms that can address the implications of this data-driven form of governance in relation to broader social justice agendas"—a broadening that, we suggest, should engage other aspects of cyberwar as well. Before taking up this issue, however, we look at another aspect of movement tactics on the battlefields of cyberwar: the attempt to hack back.

TACTICS 2: HACKING BACK

As we have seen, the social rebellions after the great financial crash of 2008 had a strong hacker component. WikiLeaks, Anonymous, Telecomix, RedHack, and other groups were described, both by themselves and by their opponents, as waging cyberwar on state and corporate power (Townsend et al. 2010). In a wave of movements characterized by a strong ethic of horizontalism, hackers and hacker groups sometimes came close to functioning as a sort of leadership cadre, whether in the charismatic figure of Assange (who on occasion referred to himself as "a bit of a vanguard"; Assange 2012, 84) or, more paradoxically, in the general adoption of the Anonymous Guy Fawkes mask as a multipurpose icon of revolt from the streets of Egypt to the squares of New York and the parliament of Poland (Olson 2012). This militancy sprang from within a hacker culture based in (though of course never entirely restricted to) male computer programming professionals. As Gabriella Coleman (2017a) remarks, such a surge of political activity by "a socially and economically privileged group of actors" once "primarily defined by obscure tinkering and technical exploration" but now willing to engage in "forms of direct action and civil disobedience so risky that scores of hackers are currently in jail or exile" is remarkable.

Searching for the causes of hacking's politicization, Coleman (2017a) finds its basis in the technosocial features of computer coding, with its "valorization of craftiness . . . cultivation of anti-authoritarianism, and . . . fellowship around labor in free spaces." These conditions—in Marxist terms the relations of programming production—tend, she argues, to be accompanied by strong commitments to civil liberties and free speech, a technical pragmatism blended with a strong critical legalism. According to Coleman, in the early twenty-first century, a mounting series of state restrictions on freedoms of digital speech and innovative practice, including intellectual property laws and national security-state secrecy, prompted many hackers to take up "the weapons of the geek." The types of hacking involved ranged widely. However, Coleman (2017b) identifies as particularly characteristic of this wave of hacker activism the "public interest hack":

a computer infiltration for the purpose of leaking documents that will have political consequence. Rather than perpetrating a hack just for hacking's sake, as hackers have always done, the PIH is a hack that will interest the public due to the hack and the data / documents.

While such leaks have antecedents stretching back to the Pentagon Papers and well beyond, digital networks give public interest hacks a particular virulence, creating exceptional opportunities for clandestine access; bulk exfiltration; and speedy, worldwide circulation, either as online info-dumps or through the mediation of sympathetic journalists.

Many of the most famous of the public service hacks associated with the 2011 cycle of struggles involved military or intelligence secrets. Several were digital mutiny by military or intelligence personnel. Private Bradley (now Chelsea) Manning was a "35 Foxtrot intelligence analyst" with the Second Brigade Combat Team of the U.S. Tenth Mountain Division. His release to WikiLeaks of the Afghan War Diary and the Iraq War Logs, including the infamous "collateral murder" video showing the killing of journalists by American attack helicopters, was enabled by access to the U.S. Army's SIPRNet (the Secret Internet Protocol Router Network) and JWICS (the Joint Worldwide Intelligence Communications System). Edward Snowden had access to NSA secrets as a worker for private defense and intelligence contractor Booz Allen Hamilton. In addition to these insider leaks, Anonymous hacked several military or intelligence contractors, such as HBGary, Stratfor, and Booz Allen Hamilton: material gathered in these breaches served as the basis for the PM Project, initiated by Barrett Brown (2012), Anonymous participant and journalist, intended to investigate and disseminate information about "the intelligence contracting industry and what is now being increasingly termed the 'cyber–industrial complex.'"

The perpetrators paid a heavy price in terms of imprisonment and exile: in the aftermath of FBI arrests of Anonymous members, North American political hacking seemed temporarily quelled.[3] Nonetheless, public service hacking and leaking has continued. In 2017, Reality Winner, a former U.S. Air Force cryptologist and translator employed by the military contractor Pluribus International Corporation, leaked a NSA report about Russian interference in the 2016 U.S. elections and was sentenced to more than five years in jail. The Panama and Paradise Papers of 2016–17 revealed the tax evasions of the superrich (including military contractors and arms dealers). Other hacks continue to illuminate shadowy wars and cyberwars. The "drone papers" published in the online journal *Intercept,* and described as the work of an anonymous "second Snowden" (Greenberg 2015), concern U.S. strikes in Somalia and Yemen and the processes by which targets are selected and approved (Scahill 2016). WikiLeaks's 2017 "Vault 7" leak of documents from the CIA details the agency's digital capabilities to compromise cars, smart TVs, all major web browsers, and the operating systems of most smartphones, not to mention Microsoft Windows, macOS, and Linux. In the same broad category are the exploits of the celebrated hacker "Phineas Fisher," who, in 2014 and 2016, breached the private companies Gamma Group and Hacking Team, which sell spyware and intrusion tools to the military and police forces with records of brutal political repression, and divulged details of their products, internal communications, and clients. Fisher's manual and tutorial video on how to "hack back" are widely regarded as manifestos for politicized hacking (#antisec 2016; Franceschi-Bicchierai 2016).

At the same time, however, hacker activities have been criticized within movements opposed to the surveillance state and neoliberal capital. Anonymous and its signature DDoS attacks have been particularly controversial. Some digital activists support them as the equivalent of the strike or picket line, an inventive and necessary extension of the means of dissent into the cyberwar arena (Coleman 2015; Sauter 2014). Others disagree, seeing such actions as provocations that invite crackdowns and infiltration: it was a turncoat Anon FBI informant who crippled the group in 2012. Evgeny Morozov moved from early support of Anonymous to declaring its libertarian support of a "free internet" too amorphous for politically coherent digital politics (Morozov 2010, 2012). Ron Deibert of Toronto's Citizen Lab notes the ease with which intelligence services might "seed an AnonOp," using the organization's facelessness as cover for its own purposes (Deibert 2013, 249).

Such problems are highlighted in Robert Tynes's (2017) account of Anonymous's combat with ISIS and the role of its offshoot, GhostSec. Following the Charlie Hebdo attack in Paris, Anonymous dramatically "declared war" on ISIS. A major umbrella group for participating Anons, GhostSec, undertook a campaign taking down ISIS websites and disabling its Twitter accounts. GhostSec at first ran as a typical "AnonOp," that is to say, anonymously, autonomously from other groups and institutions, and with information freely shared among participants. However, some GhostSec participants developed relations with U.S. operatives also battling ISIS. According to Tynes, a private security contractor, Kronos, became a bridge between the Anons and U.S. intelligence operations. Eventually, GhostSec split, as some of its members went to work for Kronos, renaming themselves "Ghost Security Group," and, "eaten by the market," commodified their anti-ISIS information for privatized paramilitary activities (Tynes 2017). The remainder of the original GhostSec persisted for a while, but then more participants entered into a relation with yet another U.S. military contractor, BlackOps Cyber, and in effect merged with it. With this, Tynes says, GhostSec dropped its Anonymous mask to become "part of a CyberHUMINT [human intelligence] counter-terrorism team."

Other problems have arisen over hacker accountability. The influential anarchist collective the Invisible Committee has objected to the use, without consultation, of their manifesto, *The Coming Insurrection,* as an on-screen message accompanying the hack of the intelligence contractor Strafor by LulzSec, a rival of Anonymous. They describe as "catastrophic" a situation where "attacks that are so political" can be "reduced by the police to some private crime" and broadly criticize the libertarian ethics of individual freedom they see pervading hacker communities (Invisible Committee 2015, 128–29). WikiLeaks's release of the "Erdoğan email," a cache of documents purportedly revealing secrets of the Turkish prime minister and his ruling party released in 2016, was strongly criticized by activist journalist and author Zeynep Tufekci (2016). She took issue not only with its inflated claims about the political significance of the hack but also with its inclusion of links to online sites containing the personal data and private emails of Turkish citizens, thereby helping to "dox most of Turkey's adult female population" at a time when activist women are often threatened with violence (Singal 2016). Tufekci (2016) remarked that while she had written in the past of the dangers of "unaccountable, centralized powers on the internet acting as censors," she "should have added another threat: unaccountable groups irresponsibly depriving ordinary people the right to privacy and safety."

Controversy also surrounds the role of Julian Assange and WikiLeaks in the 2016 U.S. election and their involvement in the release of DNC emails. The origin of this leak remains uncertain but is widely held to have been a hack by Russian intelligence operatives attempting to assist Trump's candidacy. Assange both denies receiving the emails from Russian sources and defends their release as a truth-telling exposé of political scandal, arguing that his mission is "to expose injustice, not to provide an even-handed record of events" (Khatchadourian 2017). His critics, however, suggest that he was driven by animosity to Clinton bred from her persecution of WikiLeaks while secretary of state and that his position disingenuously ignores the leak's asymmetric effects (WikiLeaks revealed no comparable dirt about Trump) and the likely involvement with an authoritarian regime's secret services. Though these accusations initially seemed flimsy, they became more serious with the discovery that Assange and WikiLeaks staff strategized with the Trump campaign over the release of the emails (Ioffe 2017; Cadwalladr 2017b). This has brought a furious response from the hacker Barrett Brown, who earlier served prison time on charges stemming from his defense of Assange but now denounced him for having "played" and "compromised" supporters (Mackey 2017).

None of this eclipses the frontline role hackers have played in exposing and defying the new cybermilitary complexes. Disagreements among political allies over tactics are widespread and by no means particular to hacker groups. The necessarily secret and risky nature of hacking makes problems of accountability almost unavoidable. However, as the Invisible Committee suggests, these are heightened by the libertarian and individualist basis of hacker culture and a reputational economy of dramatic exploits. "Hacking back" can itself be hacked back as a relay and receptor of state or parastate cyberwar. The idea that those of "the left" are preimmunized against such influence or misrecognition must be abandoned in favor of vigilant and collective autocritique, of the sort that can only arise from wider and deeper movements against cyberwar, whose prospects we review in the next sections.

OPERATIONAL

In military terminology, "operational" thinking falls in an intermediate zone between "tactical" conduct of battles and "strategic" planning of overall war aims and deals with linked activities across theaters or fronts.[4] By analogy, we term "operational" campaigns that link the dangerous and damaging effects of cyberwar to other distinct yet related points of social antagonism and political struggle within capitalism. Thus anti-surveillance activism rises from the tactical to the operational level when it moves from purely technical countermeasures to contest the legal and institutional structures of the surveillant state. In this section, we address three other issues that call for a similar approach: "fake news," cybercrime, and dirty wars.

Fake news. *Fake news* is the term now popularly used to refer to the internet's infestation with deceptive, mendacious reporting and clandestine political advertising (categories conceptually distinct but in practice often crossing-over). In North America, the issue seized public attention after the 2016 U.S. presidential election. In Western Europe, similar alarms about "fake news" have been raised in the aftermath of the Brexit campaign and French and German parliamentary elections. However, the fake-news crisis exceeds the imperial zones where its discovery has caused most uproar. Extreme and dramatic instances saturate the media of wartime Ukraine and Syria. Fake news circulates prolifically in India, much of it on behalf of the ruling Bharatiya Janata Party and its right-wing Hindu nationalist agenda (Doshi 2017); in South Africa, where Jacob Zuma's regime used it effectively against a mounting opposition; and in Myanmar, where it has contributed to whipping up atrocities against the Rohinga minority (*Economist* 2017b). This suggests that, behind the particular interests in play in the generation of online political lies, there is another question, namely, the structuring of social media that enables and creates the preconditions for fake news and the cyberwar propaganda of which it is often part.

The stage for fake news was set by a protracted crisis of mainstream corporate journalism unfolding in many parts of the world in recent decades. The combined effects of increased concentration of ownership of newspapers, radio, and television; intense spin control by state, capital, and establishment political parties; and draconian cutbacks to investigative journalism as advertising revenues drain to digital platforms have seriously reduced the quality of news reporting. Journalism has been increasingly turned into a competitive game where meticulous and rigorous investigation is overwhelmed by the search for attention-grabbing content giving ascendancy over the 24/7 news cycle (Peirano 2018). In the United States, this process reached its nadir with major media corporations' uncritical acceptance of the Bush administration's "weapons of mass destruction" rationale for the invasion of Iraq in 2003, a complicity with state power that fulfilled all the predictions of Edward Herman and Noam Chomsky's (2002) "propaganda model" of liberal capitalist media. In that sense, Trump's riposte to the charge that his election campaign ran on internet lies, a counteraccusation that legacy media were the real "fake news," contains a grain of truth.

What Trump's response willfully obscures, however, is the changed conditions of digital news propagation that favored his victory. Today, while capital's "manufacture of consent" continues to operate through the molar sites of big media companies, particularly those controlling large television networks, operations of ideological persuasion have become molecularized via the accelerated and networked circulation of social media postings and search engine results. This is occurring in a way that does not favor any true individualization of media experience, as libertarians—overlooking, or rather denying, the fast convergence of state, corporate, and military power (Glaser 2018)—had hoped. Rather, without explicit provision for a deep, equalitarian democratization of technological access and knowledge, it mainly strengthens those powers with access to the latest in profiling techniques and meme design. This includes not only large digital corporations and political parties but also previously fringe groups recruited and funded by the most reactionary fractions of capital as the shock troops of extreme marketization and social repression, such as Trump's supporters from the alt-right. It also, ironically, opens the door to digital interventions in the domestic politics of nation-states by foreign cyberwar propaganda campaigns and thence to security-state counterattacks against such digital invasions that ultimately strengthen the homeland military and intelligence apparatuses.

The major organs of news circulation are rapidly becoming social media and search, both advertising-driven businesses with revenues dependent on their ability to attract views. These corporations have, at least prior to the fake-news crisis, disavowed any editorial responsibility for truthfulness of the content they relay to their users. In the United States and elsewhere, they have also been free from regulations governing the identification of advertisers applied to older legacy media. Organization of the content delivered to users, in terms of either the ranking of search engine results and their accompanying advertisements or the supply of social media "feed," and its accompanying advertisements, is the result of black-boxed proprietorial algorithms whose operations are opaque to users and to the public. Because the algorithms are designed to keep users on-site and online, they tend to feed subscribers content they will like, regardless of veracity, and construct online communities with echo chamber and filter bubble effects. This political economy directs media technologies that, because of the immediate feedback effect of online social interactions, have exceptional attention-grabbing and -holding qualities. Thus, as Evgeny Morozov (2017) observes, "the problem is not fake news but a digital capitalism that makes it profitable to produce false but click-worthy stories."

Although social media fake news has exploded around the planet, corporate acknowledgment of the problem only occurred when it afflicted the imperial center in the 2016 U.S. presidential election. After an almost yearlong denial of the issue, Facebook finally conducted an assessment of "information operations" on its platform and presented a series of ever-expanding estimates of the number of subscribers exposed to fabricated news stories and "false amplification" (Weedon, Nuland, and Stamos 2017) processes (i.e., the industrial generation of "likes"). Google and Twitter have made similar assessments. The distinct but related issue of advertisements from various officially suspect sources—including revolutionary jihadists and racist groups—also became a scandal. The response of search and social media giants has been to promise greater screening, by both thousands of human moderators and artificial intelligence automation, and adjustments to their algorithms to exclude such content.

This, however, has only opened further questions about the large digital corporations' already vast powers. Discussion of intensified content moderation has brought to the surface the censorial powers they already exercise over social media content, in relation to both depictions of violence or sexuality and political topics. There is a very real likelihood that the blocking of fake news will become a mandate for the suppression of any politically volatile internet content deemed problematic by state authorities—a solution tantamount to the adoption by liberal capitalism of a "Great Firewall" approach to feared foreign influences and domestic dissent. Twitter's liquidation of suspected Russian bots has elicited outrage from the alt-right, some of whose active online celebrities claim they have lost hundreds of followers (S. Gallagher 2018c). But corporate tweaking of search and social algorithms is also claimed to have brought sharp declines in internet traffic to left-wing websites (Damon and North 2017; Damon 2017). Some of the issues at stake are highly technical, but it's precisely the proprietorial opacity of the "algorithmic ideology" (Mager 2012) exercised by social media and search engine capital that is now thrust into visibility.

There is thus an important connection between opposition to escalating cyberwar and critique of the digital oligopolies that have made search and social media instruments of antisocial disintegration. Even within the U.S. high-technology community, famously opposed to state regulation, voices have been raised expressing skepticism as to the capacities of Google and Facebook to police themselves, calling for enforced transparency and even financial liability for the "negative externalities" of their algorithmic processes (Wu 2017; Parakilas 2018). Beyond these criticism are, however, deeper ones—for example, the degree to which capital's media systems are deeply and basically dependent on the "fake news" of advertising and on the "false amplification" (Weedon, Nuland, and Stamos 2017) of highly engineered attention manipulation. The 2018 disclosure of Cambridge Analytica's secret acquisition of some 87 million Facebook users' personal data—data whose covert collection in 2014 was at that time perfectly permissible according to the platform's terms of service—provoked a brief crisis for the giant social media company. Its shares fell; Mark Zuckerberg was summoned to U.S. congressional hearings; a "delete Facebook" movement sprang up. However, this moment of outrage was rapidly recuperated. Zuckerberg promised reforms even as his company in actuality persisted in evading privacy laws (BBC 2018a) and expanding the scope of its intrusions with controversial facial recognition technologies (BBC 2018d); Facebook's financial position rebounded; and a flurry of discussion over government regulation of social media subsided. Whether this outburst of user revulsion against mass surveillance was only a flash in the pan or a foretaste of a more sustained wave of disaffection remains to be seen.

Morozov (2017) writes,

> The only solution to the problem of fake news is to completely rethink the fundamentals of digital capitalism. We need to make online advertising—and its destructive click-and-share drive—less central to how we live, work and communicate. At the same time, we need to delegate more decision-making power to citizens—rather than the easily corruptible experts and venal corporations.

The issue of cyberwar thus connects to the need for radical thinking about decentralized public control over the means of communications, extending beyond the regulation or breakup of large social media and search engine oligopolies to the establishment of "commons" institutions separated from both state and market; the formation of public computing utilities kept at arm's length from government, and "a socialization of the data-banks," without subjecting networks to more direct censorship by national security apparatuses (Morozov 2015). This coincides with a wave of concerns about, and, in Europe, litigation against, the giants of digital capitalism over violations of public regulation and labor law; tax evasion and anticompetitive abuse of monopolistic powers; and, more generally, the control of socially transformative technological research and innovation exercised by the giants of information capital (Foer 2017; Mosco 2017; Taplin 2017; Wu 2017).

Cybercrime. Given how states criminalize hacking against corporate or military interests, it is hazardous to invoke the concept of "crime" in a critique of cyberwar. Yes, it is worth the risk, for just as there are war crimes, so there are cyberwar crimes. In his history of the U.S. national security state, Alfred McCoy (2017) details the "covert netherworld" in which military and intelligence agencies collaborate with criminal networks for mutually beneficial purposes around government corruption, drug trafficking, the arms trade, and other black markets:

> While the illegality of their commerce forces criminal syndicates to conceal their activities, associates, and profits, political necessity similarly dictates that secret services practice a parallel tradecraft of untraceable finances, concealed identities, and covert methods.

The personnel of both worlds inhabit "clandestine milieus" that often overlap (McCoy 2017). Such relationships are of course not unique to the United States. Other nation-states engage in similar practices—Russia's crime rings have played a significant role both in the reestablishment of capitalism in that country and in its security service operations.

Cyberwar has created a whole new state-criminal netherworld of "cybermercenaries" (Maurer 2018). As we have seen, one of the multiple trajectories that emerged from this original hacker matrix was for-profit computer crime. The pursuit of cybercriminals became an objective of both state law enforcement and private cybersecurity firms. Nonetheless (and despite the lack of any truly reliable indices), it is likely that over the last three decades, the scale of cybercrime has grown due to increased access to computers and networks; economic globalization; the advent of social media, yielding rich hauls of personal information and passwords; and technological leaps in the automation of network penetration or destruction. Very large-scale data breaches, such as those that affected Yahoo!, Equifax, Ashley Madison, and Uber in 2015–17, are now regular features in the landscape of digital capitalism, while ransomware attacks have become a scourge for both large institutions and individual users.

At the same time, however, military and spy services adapting to the new "clandestine milieu" of digital crime draw on the skills and knowledge of so-called black-hat hackers to infiltrate, steal from, and sabotage opponents' networks. Indeed, for states aspiring to be cyberwar powers, "it is by no means necessary to have the logistics base in house," because

> cyber-crime actors can sometimes be a vital part of this equation, developing cyber capabilities for their state sponsors that are first used by the state actors but are later released back to the cyber-crime

ecosystem when they are not considered state of the art anymore. (Klimburg 2017, 305)

By definition, this "clandestine milieu" is hard to look into; however, of late, a few windows have opened to illuminate its murky depths. Thus, in his account of the rise of the U.S. "military–internet complex," investigative journalist Shane Harris (2014) describes the NSA's involvement in the zero-day market. *Zero days* are security flaws or bugs in software programs—usually new releases or updates—that the provider is not aware of: zero day is the day the provider becomes aware of the vulnerability. Hackers seek out zero days, often by intensive research, because, for as long as the corporate producer remains ignorant of, or fails to fix, the flaw, they offer ways of illegally entering or compromising networks. If they involve widely used off-the-shelf programs, such as those of Microsoft, iOS, or Oracle, the vulnerability can be of mass scale. The hacker who discovers a zero day may either exploit it herself or sell the information to others; vending zero days has become a commercial activity on the dark web, where it is one component of a thriving cyberarms trade that also includes malware sales and botnet rentals.

In 2016, following a Freedom of Information Act request by the Electronic Frontier Foundation, the U.S. government revealed that under its "Vulnerabilities Equities Process," a consulting group of security and law enforcement agencies determined on a case-by-case basis how to treat zero days of which they became aware. The decision was whether to disclose them to the public or keep them secret for offensive use against enemies. The NSA itself became a major purchaser of zero days, building an arsenal of known network vulnerabilities that could be used against adversaries—a purchasing program that both legitimized and incentivized zero-day markets. Harris (2014, 95) reports that, with a $25 million 2012 budget, the NSA was at one time "the single largest procurer of zero-day exploits," many of which it obtained with the help of major American defense contractors Raytheon and Harris Corporation, which played an important role as "middlemen" in the zero-day market. The NSA is believed to have stockpiled "more than two thousand zero-day exploits for use against Chinese targets alone" (Harris 2014, 96). In the wake of the Snowden revelations, NSA acquisition of zero days was officially restricted, though questions continued about both exceptions to the ban and the trove that it had already collected. In 2016, a high-level NSA hacker declared that the agency could get along quite well without zero days because "there's so many more vectors that are easier, less risky and quite often more productive than going down that route" (Kopfstein 2016).

This was, however, not the end of disclosures about the interface between cybercrime and cyberwar. The Equation Group is a highly skilled hacker group generally believed to be associated with the NSA. In 2016, another hacker group, the Shadow Brokers, suspected of a Russian connection, broke into an online Equation Group cyberweapons depository and stole hacking tools from it, including several zero-day exploits targeting enterprise firewalls and antivirus software; of particular importance was to be EternalBlue, a vulnerability to Microsoft Windows. After attempting to auction the weapons cache online, the Shadow Brokers eventually distributed them for free, as a sort of cybercrime potlatch, possibly to discredit and embarrass the NSA.[5] Subsequently, items from the cyberweapons cache have been used in criminal cyberattacks.

In particular, the EternalBlue vulnerability featured in the very large-scale ransomware attacks in 2017, including the WannaCry virus that reportedly infected more than 230,000 computers in over 150 countries, temporarily crippled parts of the United Kingdom's National Health Service, and also affected Spain's major telecommunications company. Many sources declared that at least part of the blame rested with U.S. intelligence services. Microsoft president and chief legal officer Brad Smith (2017) wrote,

> Repeatedly, exploits in the hands of governments have leaked into the public domain and caused widespread damage. An equivalent scenario with conventional weapons would be the U.S. military having some of its Tomahawk missiles stolen.

WannaCry seemed at first to have been a commercial crime. However, in 2017, the British government asserted that it was "almost certain" it was the work of North Korean Lazarus Group, formerly associated with hacks against Sony and a cyberheist of Bangladesh's national bank: this is believed to be an important factor influencing a draft European Union policy statement that serious cyberattacks can be considered acts of war and asserting that individual member states could "in grave instances" respond with conventional weapons (Knapton 2017).

The episode presents a situation in which three major hacker groups, each of which has an alleged association with a nation-state, are implicated in a worldwide network breakdown of significant proportions. While WannaCry was a malware infection of exceptional scope, it is far from unique; other ransomware attacks have affected several regions. As early as 2012, it was reported that zero-day exploits incorporated into the Stuxnet worm, used in the U.S.–Israeli attack on Iran's nuclear reactors, had been picked up and used widely by cybercriminals, who were also copying sophisticated design elements of weapons, such as forged Microsoft security certificates (Simonite 2012). Other forms of attack, such as the Mirai botnet, which effectively turned the huge amounts of data generated by video surveillance systems into a means for disabling websites, have also intensified in scale. As Tim Maurer (2018, 161) observes, distinctions between "crimeware" and "milware," that is, between criminal malware and military malware, are becoming "less meaningful." Of special concern to some cybersecurity experts is that some attacks that appear initially to be extortion attempts are in fact "wiper" attacks that destroy their targets with the possibility of ransomed restoration—leading them to believe that, rather than commercial criminal attacks, these may be tests for attacks simply intended to destroy network capacity (Schneier 2016). Cybercrime is an ambiguous issue. Its threat is frequently invoked to justify enhancements in the powers and budgets of police and spy agencies and to encourage the sale of cybersecurity products. However, the realization that security establishments themselves are involved in, and exacerbate, criminal enterprise opens the way to a progressive appropriation of law-and-order issues that would have a precisely contrary direction and argue for the deescalation and decommissioning of cybernetic armaments.

Dirty wars. Cyberwar fosters fake news and digital crimes, but above all, it fosters war. Its damage cannot be estimated simply in terms of deceived voters, frustrated internet users, or even canceled medical operations (in hospitals crippled by malware), freezing homes (amid electrical blackouts), or destroyed centrifuges (in sabotaged nuclear plants). It includes now the dead, maimed, and wounded. We have stressed the difficulty, indeed, the impossibility, of segregating cyberwar as a distinct domain of military activity; on the contrary, we have argued that it is characterized by a tendency, intrinsic to the constantly growing scope of the digitization of all spheres of life, to overspill boundaries. And nowhere is that truer than in the realm of war itself, where, as a recent article in a U.S. defense journal put it, "'Cyber War' Is Quickly Becoming Just 'War'" (Tucker 2017). Or, as we have put it already, "hybrid wars" that synthesize cybernetic and kinetic elements are today becoming the norm. This means that opposition to cyberwar is inseparable from resistance to other types of war.

Here we will look at how this cyberlogic plays out in in the relation to *dirty wars* (Scahill 2013), also known as small wars, irregular wars, shadow wars, or ghost wars. These are the wars without official declarations of hostilities, at and beyond the boundaries of international law, that today proliferate as counterinsurgency operations or proxy conflicts antagonizing different populations against each other. Such wars now involve advanced digital systems for intelligence gathering and analysis, coordination, and weapons delivery, functions fused in weapon platforms that are also integral parts of cyberwar. Thus among the first, and most representative, victims of cyberwar can be numbered the populations of areas in Pakistan, Afghanistan, and Yemen living under visible and audible overflight by CIA Predator and Reaper drones, from which death suddenly and unpredictably descends.

Drone warfare has been widely discussed (Turse 2012; Benjamin 2013; Chamayou 2015; Cockburn 2015; Scahill 2016; Shaw 2016), but we want here to put it in the wider context of digitized and networked conflict. As semiautonomous flying vehicles, drones push toward the full robotization that was always a goal of cybernetics. Their remote operations depend on streams of digitized information, flowing from the Nevada bases that house their video game–trained "pilots" to satellite relays in Germany, boosting the signals onward to the aeronautical entities roaming the heavens of the Middle East, entities whose navigation is a matter of computerized geolocation and which are targeted largely on the basis of electronic signals intelligence (SIGINT). In their overflights, U.S. drones are not merely surveilling territory and attacking targets; they are also sucking up vast quantities of wireless data, primarily cell phone signals, which can be collected through machines like the Gilgamesh device, which can be attached to the base of the drone and operates as a "fake cell phone tower," forcing targeted SIM cards, purportedly in phones belonging to suspect terrorists, to contact it (Scahill 2016, 66). Information gathered in this way is relayed back to data fusion centers for combination and cross-matching with other intelligence sources.

How much of this mix comes from HUMINT and how much from SIGINT depends on the theater of operations. In areas like Pakistan, where the United States has a significant number of special forces and informants, the human component may be quite large; in others, such as Somalia, where it has very few operatives, that dwindles. In either situation, electronic data, especially from mobile phones, are likely to play an important role in "find, fix, finish" operations. "Most drone strikes are aimed at phones" (Scahill 2016, 83). This is the case both for "personal strikes," in which the identity of the hunted person is known, and "signature strikes," where the target simply exhibits a suspicious form of behavior.

In the case of personal strikes, the identity of the subject is in fact likely to be established by the interception of communication signals, notably cell phone signals, associated with the victim: drone operatives are not tracking the person but the phone that is purportedly attached to the person. This probabilistic element is even more intense in the case of signature strikes, where the trigger is a departure from or conformity to certain "pattern of life" configurations, which are built up by a combination of multiple data sources; "imagine a superimposition, on a single map, of Facebook, Google maps and an Outlook calendar . . . a fusion of social, spatial and temporal particulars . . . a combination of the three dimensions that, not only in their regularities but also in their discordances, constitute a human life" (Chamayou 2013, 48). This is "death by metadata" (Scahill 2016, 94; see also Cheney-Lippold 2017, 39–47). It is also a process that, because of its reliance on algorithmic calculation, tends toward the creation of what Chamayou terms "political automata," whereby life and death are cybernetically determined.

Drone attacks exist in a gray zone between war, assassination, and death sentence; they are executions with collateral damage. While the numbers of deaths attributed to the U.S. drone campaigns is contested, it is widely reckoned at between three and four thousand and will be higher once their use in the air war against ISIS in Iraq and Syria is reckoned in. While the official estimates of civilian casualties is low, critics suggest that this is because of a near-automatic designation of adult male deaths as those of enemy combatants. There have been several dramatic cases of misidentification by SIGINT. It appears that, in particular areas and time periods for which data are available, some 90 percent of deaths by drone may be people other than the specified targets (Scahill 2016). Because of such issues, drone warfare is unpopular on a worldwide basis—except within the nations waging it. Despite the protest of jurists, journalists, humanitarians, and disaffected drone pilots, drone strikes continue to enjoy popular support in the United States as a means of fighting the war on terror without risking U.S. casualties (such support falls sharply in incidents when Americans are accidentally killed). It therefore seems implausible to suggest that opposition to drone attacks and other forms of "black ops" could become a contributory part of a social resistance to cyberwar.

However, as several observers have noted, the popularity of war drones in the United States is contingent on their perceived success and on the monopoly of their use. There are serious military doubts about the long-term prospects of an assassination policy that alienates the regions subjected to it and, in its repetitive "mowing the grass" of terror suspects, seems part of a recipe for endless war. Even more to the point is the obvious fact that the current imperial franchise over drone technology is temporary, as witnessed by events like the successful remote hijacking and capture of a U.S. drone by an Iranian cyberwarfare unit in 2011 and by the tactical use of "homemade" drones by ISIS during the battle for Mosul. As the danger of other people's drones intensifies, their glamour will likely decline. In this respect, the drones and their kin will follow the broader path of mutual escalation charted by all other technologies of cyberwar and, as they do so, may be more widely opposed to aspects of a generalized and potentially disastrous militarization. In this regard, the events around Google's decision to provide artificial intelligence to Project Maven, a military program to accelerate analysis of drone footage by automatically classifying images of objects and people, are of major significance. In March 2018, nearly four thousand Google employees signed a petition objecting to their company's involvement; three months later, eleven of them resigned in protest (Conger 2018b). Google then canceled the project, an extraordinary victory for opponents of digital militarism.

ORGANIZATIONAL

So far we have suggested how opposition to cyberwar could emerge as part of wider movements contesting a capitalism whose globally totalizing yet nationally competitive logic drives the militarization of networks. We now make a brief digression to consider issues of political organization that bear on the formation of countermovements against the new wartimes. In particular, we look at some implications of cyberwar for the controversy about parties and networks, verticalism and horizontalism.

The "vanguard party," the form of political organization most famously (or notoriously) associated with Marxism, has an obvious military reference. This martial derivation is implicit in the metaphors Marx and Engels ([1848] 1964) use in *The Communist Manifesto* to describe the party leading the "line of march" of proletarian movements. Such metaphors were grounded in the actuality of mid-nineteenth-century revolutionary uprisings, whose military aspects deeply interested Engels. It was, however, in the context of Russian Marxism that Lenin ([1902] 1969), an enthusiastic reader of Clausewitz, fully articulated the latent concept of the vanguard party (Kipp 1985; Boucher 2017). This party was forged in a context of revolutionary terrorism and Tsarist counterinsurgency. Then, in the midst of world war, it seized power by armed insurrection and held it through civil war and foreign invasion. The authoritarian aspects of militarized vanguardism were, from their inception, denounced not just by anarchists but also by the Marxist ultraleft, but Bolshevik victory gave it the sanction of success. Subsequently, in other times and regions, vanguardism would adapt to contexts very different from the European conflicts in which it originated, as, for example, in Maoist concepts of "people's war" (Mao [1938] 1967). Yet, however remodulated, the concept of the party as quasi-military formation leading progressive advance, and, if necessary, fighting for it under actual war conditions, remained a defining element of twentieth-century Marxist imagination and practice.

After the fall or retreat of state socialist regimes in 1989, vanguardism was widely discredited. In the aftermath of world-historical defeat, digital networks encouraged the concept of a multiplicitous and horizontal left whose self-organization would be achieved without special leadership cadres. Some proposing this path were entirely antipathetic to the armed struggle lineage of Leninism. For others, the issue was not so much a disavowal of military models (at least at the metaphoric level) as a resort to hopes for a different form of war, waged without centralized command. Such ideas figured largely among autonomists, including one of the authors of this book (Dyer-Witheford 1999), and other ultraleft Marxists associated with turn-of-the-century alter-globalism, with its independent media centers, hacktivism, and electronic civil disobedience. The subsequent adoption of network-centric warfare by state powers, however, relays back to anticapitalist movements lessons more complex than any straightforward vindication of horizontalism over verticalism, network over party. For what today's cyberwars demonstrate is that though networks complicate hierarchy, they are not antithetical to it. The Pentagon, Kremlin, and other state cyberwar practitioners now integrate hacker proxies, digital militias, and semiautonomous special operations units within top-down command strategies. Indeed, sometimes these networked collocations generate new vertical chains of command.

To take a controversial example, consider Steve Niva's (2013) account of "the Pentagon's new cartography of networked warfare." Although U.S. military intellectuals had for years been discussing "network-centric" warfare, Niva says practical steps toward this occurred only through military adaptation to the problems in the field arising from its invasions of Afghanistan and Iraq. The crucial moment, he claims, was the creation of the Joint Special Operations Command (JSOC) to fight a Sunni insurgency bringing the U.S. occupation of Iraq to the brink of defeat.

Under the auspices of General Stanley McChrystal, who took as his motto "it takes a network to fight a network," JSOC, constituted initially as a specialized commando strike force, was built up as a command linking different branches of the regular U.S. military, its special operations units, the CIA, and the NSA in actions conducted on the basis of "common information and self-synchronization." The coordination of drone strikes, commando raids, armored snatch squads, databanks, and signal intercepts into a devastating "war machine" raised counterinsurgency warfare to an industrial scale and inflicted a crushing (though ISIS would later show, temporary) defeat on the rebellion. It was "hybrid blends of hierarchies and networks" that made JSOC into an "organizational hub and revolutionary motor" of warfare across the U.S. military, so that, "as both incubator and example," it became the leading force in the U.S. "war on terror" (Niva 2013, 185). In short, JSOC is what a vanguard looks like today. And it also looks a lot like a network.

Let us now put this beside a very different discussion of the intermingling and coexistence of vanguards with networks. In his *Organisation of the Organisationless: Collective Action after Networks*, Rodrigo Nunes (2014b) reviews "lessons learned" from the 2011 "take the square" movements in Cairo, New York, and Madrid, in whose eruption digital communication played an important role. As he points out, contra popular media depictions, such movements were never manifestations of pure horizontal spontaneity and networked contagion. On the contrary, though much of their organization did indeed take place on networks, certain core nodes "with an anomalously high number of links and with links to nodes in more and more distant clusters" were of exceptional importance in that initiating calls to action and attracting followers were clearly apparent (Nunes 2014a, 175). These nodes performed what Nunes terms the *vanguard-function*. However, any given node only lasted for a limited period of time; nodes changed over time as the movements went through different phases of action, so that their rotation involved a form of "distributed leadership." Tahir Square, Puerto del Sol, and Zuccotti Park are what networked revolution looks like. And it includes "vanguard functions."

For Nunes, the persistence of a form of verticality in these often self-declaredly horizontal movements is not a flaw in their architecture.

Rather, the problem of the 2011 movements was rather their reluctance to recognize the necessary constitutive function of this verticality within movements' self-proclaimed horizontality and hence to work constructively with it. The relative failure of both alter-globalization "summit busting" and 2011 "take the square" movements, both of which relied heavily on a digital circulation of struggles, has led to revived debates about the need for centralized party organization that would provide greater durability, discipline, and long-term orientation of movements. But, to paraphrase Nunes, any kind of contemporary party organization would have to emerge from within the network setting and hence be unlike classic vanguardism. It would be a party of a networked type—and hence probably, we would add, requiring a new name too, as the left remembers the long heritage of forms that lie somewhere between spontaneity and vanguardism, horizontality and verticalism, in leagues, federations, and fronts.[6]

It can be objected that the parallelism we have so far suggested between the cyberwar organization of state and revolution is misleading. For even if state war fighting now occurs within an organizationally complex field assembled via networked forms, it is ultimately subservient to command from on high (just as in post-Fordist capitalism, the institution of work teams responsibilizes employees only within managerial limits). In contrast, all but the most Stalinist concepts of vanguardism have in their background some acknowledgment, however unfulfilled, that the party is an instrument of proletarian will. Chris Hables Gray and Ángel J. Gordo (2014, 251) insist that "horizontalist social movements incorporate new information technologies into their praxis as self-control, while militaries seek to subsume them into the existing hierarchical control paradigms." We agree that the relation is not isomorphic. Our point is that the art of cyberwar is a game not of horizontals against verticals but of diagonal permutations. In this game, the play of the security state and the forces of rebellion are neither fully identical nor completely incommensurate. Rather, they are inversions of a shared problematic. The play of the state is to incorporate horizontal forms while subordinating them to vertical lines of command, the gambit of revolution to allow and include moments of vertical decision subordinated to a horizontal accountability.

With this in mind, our final example of organization in a cyberwar context returns to the virtual and physical battlefields of the Middle East and the Rojava Revolution—that is, the wartime experiment in new forms of social collectivity, undertaken by Kurdish communities in Northern Syria, involving direct democracy, female emancipation, and environmental care. These experiments developed out of the long struggle for the establishment of a Kurdish nation waged by the Marxist Kurdish Workers Party, a Leninist vanguard organization conducting armed resistance against the Turkish state. Over the last decade, however, this movement has, under the influence of its imprisoned leader Abdullah Ocalan, undergone a doctrinal transformation in which Marxism is resynthesized with anarchist, feminist, and ecological thought, as well as Kurdish culture, to produce a new concept of revolution from below. It is a metamorphosis of Marxism in some way similar to that undertaken by the Zapatistas in Chiapas a decade earlier, albeit in a very different context. This is not the place to discuss the complexities and contradictions of the Rojava Revolution, now a topic of wide discussion (Knapp 2016; Küçük and Özselçuk 2016; Üstündağ 2016). We will just make three points.

First, perhaps the most prominent feature of the Rojava Revolution is its combination of vertical and horizontal elements. A central aspect of Ocalan's rethinking of revolutionary doctrine, and encounter with anarchism (mainly through the works of Murray Bookchin), has been a rediscovery of the antistatist elements of Marxism and a critique of the despotic tendencies of state-governing elites. As David Graeber (2016, xvi), one of the most eloquent supporters of Rojava, explains, what this means in practice is the development of a form of "dual power," not divided between the revolution movement and the state it attempts to overthrow (as in Leninism) but rather within the revolution itself. The doctrine of *democratic confederacy* (or sometimes democratic autonomy) involves two apparently contradictory elements. On one hand, Rojava's regions or cantons are organized by a series of apparently conventional formal governing institutions: ministries, parliaments, and courts. On the other, however, there is a set of horizontal popular assemblies, communes, councils, and assemblies. The former are meant to be accountable to the latter. As Graeber points out, this "integration of top-down and bottom-up structures" (xx), in which the same people often occupy positions on both vertical and horizontal axes, is complex, and sometimes confusing, but it does represent an attempt to resolve some fossilized conundrums of the left.

Second, the Rojava Revolution is made in conditions of war—and cyberwar. It is a social laboratory constructed amid the catastrophe of Syria's internecine conflict, attacked by ISIS, threatened by Turkey, with sporadic military support from the United States. Its war involves mainly fierce conventional fighting, as in the epic defense of Kobane against ISIS, but also cyberwar elements. Kurdish struggles, because of their diasporic elements, have always depended on transnational communication, and recently digital networks and social media, persistently attacked by enemies such as ISIS and the Turkish state (Neubauer 2017). Like other parties in the Syrian conflict, Rojava mobilized support and recruited foreign fighters via virtual sites, such as through the Lions of Rojava Facebook page (Johnson 2015). As in all hybrid wars, digital communication is important to its battlefield operations, probably including the targeting of supporting U.S. air strikes against ISIS (a matter of immense controversy on the left). It has also attracted support from hackers. In 2016, Phineas Fisher reportedly donated the proceeds of a bank cyberheist in the form of twenty-five bitcoins (then worth approximately US$11,000) to support Rojava, which he described as "one of the most inspiring revolutionary projects in the world" (Paganini 2016), while anarchist cryptocurrency experts have also attempted to help the Rojava economy (Greenberg 2017a). The Rojava Revolution's organizational inventiveness certainly can't be derived from its digital elements, but it emerges in the context of the most contemporary techniques of digital rebellion.

This in turn relates to our third point. Rojava has its genesis in Leninist armed struggle. But Ocalan's metamorphosis of revolutionary thinking was in large part a critique of the separation of vanguard militants from the populations that support them. Üstündağ (2016) argues that the Rojava Revolution has occurred in the historical context of a movement, under attack from the Turkish military and, even more murderously, ISIS, where "nonviolence" is not an option. The radicalism in its rethinking of military vanguardism lies in development of "self-defense" capacities, not as separated state functions but as organizations rooted in communities and seen as integral to wider social transformations. The most dramatic aspect of this is the role of women fighters, which has been central to the struggle against ISIS but also challenges the internal patriarchy of the revolution. The People's Protection Units and Women's Protection Units—whose formation seems a recovery of the capacities for what Virilio ([1978] 1990) terms "popular defense"—are integral both to defense against murderous attacks and to wider social metamorphosis. We find Rojava inspiring, but we don't view it uncritically or pretend to understand it thoroughly. Its prospects, like those of the entire region in which it is situated, are immensely uncertain. There is also a risk that its struggles will become the basis for "phantasmal projection" of distant leftists (Küçük and Özselçuk 2016, 185). There are clearly unresolved "ambiguities" (Leezenberg 2016) in Rojava's integration of its Leninist heritage with elements of grassroots communization. Nonetheless, we find in its experiment an encouragement to consider new possibilities for revolutionary organization that go beyond the time-worn antimonies of vanguard and network, vertical and horizontal; while this discussion goes well beyond issues of cyberwar, it is one to which the study of cyberwar can contribute.

STRATEGIC

Manolo Monereo (2017) writes, "Historians agree that between 1871 and 1900 there was a first capitalist globalization, which, it is worth emphasizing, ended in what has been called the 'thirty-year war.' Today we are at the end of a similar cycle." This conjunction is, Monereo suggests, composed of four basic elements: "1) recurrent economic-financial crises; 2) seismic geopolitical shifts, with a massive redistribution of world power; 3) aggravation of the planet's socio-ecological crisis; and 4) a crisis of 'occidentalism,' calling into question the entire project of Western modernity." The phase that is unfolding, he concludes, "will not be characterized by peace, stability, and free-market cosmopolitanism, but a struggle between world powers vying to carve out of zones of influence and divide up natural resources. And, regrettably, by war"—to which we would add, especially and characteristically by cyberwar.

As we have argued, from 1945 on, the hegemonic status of the United

States, as the world's chief capitalist power, was intrinsically related to the development of computers and networks. The role of digital systems in its military–industrial complex, initially tightly coupled with nuclear weapons, spread through other aspects of its war-making system as well as through the general economy. In both aspects, it contributed to the United States's eventual Cold War victory. In the aftermath of that victory, the United States continued to develop its digital military capacities into the ever more direct weaponization of network, creating the technological–human assemblages of what is today referred to as cyberwar. The scope of NSA global surveillance and sabotage programs and the sophistication of the Stuxnet nuclear centrifuge-destroying malware are only the most manifest instances of this process, which is today an integral part of a wider upgrade of U.S. military capacities that ties together a nuclear primacy with the militarization of space and drone warfare. Accompanying and spurring on this process is the additional dynamic of cyberwar adoption by the forces antagonistic to the global dominance of the United States and its allies. These antagonists include the defeated socialist powers, Russia and China, now paradoxically resurrected as capitalist competitors in the world market, or, in the case of North Korea, surviving in a macabre afterlife of state socialism. They also include the forces of militant Islamic jihadism, beckoned into existence by the West as an anticommunist ally, only to become its opponent in the long war on terror. All these actors converge on the militarization of digital networks. Many observers today see a moment that recapitulates the decline of previous imperial hegemons within the global capitalist system—Spain, Holland, Britain—and parallels the moments of extreme instability as old powers and new contenders confront each other.[7] The rise of cyberwar is part of this tumult and quite possibly a precursor and preparation for widening and intensifying conflict. Schematically, we can envisage three potentially intertwining trajectories such a process might take:

I.   **Network degradation.** Alexander Klimburg (2017) outlines the possibility of a "darkening web" characterized by persistent and gradually intensifying cyberwar between states and between states and terrorist movements conducted in a variety of registers. Security breaches, aggressive malware, and botnet attacks proliferate. Digital industrial sabotage and critical infrastructure attacks begin to multiply, as do the accidental runaway effects of cyberweapons. Networks are deeply and chronically infected with computational propaganda, fake news, and viral mis- and disinformation. In response to adversarial incursions, states intensify algorithmic surveillance, censorship, and preemptive virtual policing. Cybersecurity provisions become increasingly mandatory and elaborate. Attribution problems, falsification of evidence, and the overlap between military and intelligence forces and criminal networks create a chaotic digital twilight of hacking and trolling, botnets and viruses, malware, surveillance, and bugs, shutdowns, blocking, and filtering, in which uncertainties exacerbate suspicions and hostilities, altogether making the internet increasingly impossible to use. In short, the "darkening web" is what already exists now, only more so. One of the cofounders of Twitter, Evan Williams, offered his diagnosis, suggesting that "the Internet is broken" (Streitfeld 2017). But maybe it's not. Maybe the internet is finally what it was always meant to be. Maybe it is perfect, but not for us, the excommunicated user-subjects. For cyberwar.

2. **Hybrid escalations.** Similarly rooted in the present is the likelihood that the simultaneous virtual and kinetic conflicts, such as the Syrian civil war, the fighting in Donbas, and the many branches of the war on terror, continue and break out in new regions, bringing ever higher levels and varieties of cyberweapons, deployed for purposes ranging from intelligence gathering, battlefield surveillance, and munitions delivery to sabotage of enemies' domestic and military resources. The use of drones and other semi- or fully automated weapons systems expands and takes new directions, such as the development of swarms of small autonomous vehicles—"slaughterbots" (*Economist* 2017a)—for house-to-house fighting in ruined cities. The biometric and networked tracking of refugees created by such conflicts, and the control and interdiction of their entry to affluent fortressed homelands, becomes a major activity of the nation-state security apparatus. Because present hybrid wars are also in large part proxy wars, where local battlefield actors are directly or indirectly supported by major

powers, they are charged with the possibility of abrupt collisions between the most powerful militaries on the planet.

3. **"Thermonuclear cyberwar."** We borrow this phrase from Erik Gartzke and Jon Lindsay (2017), who are among several authors currently pointing to a renewed and dangerous rendezvous between cyber- and nuclear weaponry. The last decade of debates between defense intellectuals about cyberwar has split those who see digital attacks a new equivalent of nuclear weapons, capable of disabling whole societies through critical infrastructure attacks, and skeptics who deride such anxieties as hyperbolic and implausible. But "cyber" and "nuke" are not separate. As we have seen, they were twinned at the moment of conception, with the development of each dependent on the other. And the connection is not just historical; it is current. Now cyberwar weaponry is part of a new approach to nuclear war fighting, the left-of-launch approach. Early ventures in antiballistic missile defense, such as Reagan's "Star Wars" strategic defense initiative, depended on shooting down swarms of missiles as they plunged through the atmosphere toward their target. Left of launch, in contrast, aims to "strike an enemy missile before liftoff or during the first seconds of flight," using "cyber strikes, electronic warfare and other exotic forms of sabotage" (Broad and Sanger 2017). This doctrine was incubated during the Obama administration and inherited by the Trump presidency. Advocates of the left-of-launch nuclear strategy present it as a defensive measure. However, the doctrine destabilizes basic premises of deterrence that have, since 1945, restrained nuclear weapon use (Cimbala 2017). Deterrence depends on a dread faith by all parties that both their own and their enemies' nuclear weapons will work. The possibility that nuclear weapons systems might be secretly disabled raises prospects both of overconfidence (trusting one can sabotage an opponent's system) or panicked preemption (fearing left-of-launch attacks on one's own nukes and falling into a "use 'em or lose 'em" mind-set). More generally, control and command of nuclear weapons depend on communication systems whose collapse in a crisis situation could

have catastrophic results.[8] The origin of the internet lay in the U.S. attempt to ensure continuance of such systems in the event of nuclear war; now the weaponization of the internet itself constitutes a possible cause of nuclear war.

Facing such prospects, liberal commentators propose diplomatic measures to control and mitigate cyberhostilities. Klimburg (2017), for example, suggests a series of initiatives to be undertaken primarily by the United Nations and the Internet Corporation for Assigned Names and Numbers (ICANN), the long-standing (and controversy-ridden) forum for internet governance. In these venues, he suggests, it should be possible to work out a series of agreements—"digital arms" limitation treaties, comparable to those on nuclear weapons; an "attribution and adjudication" council to assess and arbitrate responsibility for cyberattacks; international cooperation against cybercrime; the promotion within ICANN of "civil society" perspectives to counter those of states and corporations. He suggests that "standing bodies" regulating cyberwar would be comparable to the Intergovernmental Panel on Climate Change and emphasizes the importance of "scientific and authoritative advice from experts to political decision makers on how to avoid disaster" (344).

Such proposals, seemingly eminently sensible, ignore the reality that the tensions driving the rise of cyberwar are also incapacitating the fragile apparatuses and institutions of international cooperation that have existed since 1945. As Jon Lindsay (2012) observes, while proposals for cyberwarfare treaties are "well meaning," they would, within the current state of international great power relations, be "hacked to bits," because "cyberoperations, like other types of intelligence and covert operations, take place in the shadows. An international treaty on cyberweapons would be . . . totally unenforceable, since such activity is designed to evade detection and attribution." The conjuncture in which cyberwar rises, and part of the reason for its ascent, is the breakdown of nuclear arms limitation and nonproliferation treaties. Klimburg's comparison of the regulation of cyberwar and climate change is unfortunately all too telling, given the failure of global capitalism to generate any binding interstate agreements on carbon emissions and the recent withdrawal of the United States from even the nonbinding Paris accord on global warming. And like slowing climate change, reducing the risk of cyberwar requires deep, systemic social change.

The argument of this book is that cyberwar is a manifestation of the competitive nature of capitalism, which, beneath the surface of globalization, fosters a war of all against all, conducted in the accelerated, automated, and abstracted forms on which this entire mode of production now depends. It follows from this that the prospects for reducing the dangers of cyberwar, and of the other types of war of which it is now part, depend strongly on movements and struggles to constrain and, ultimately, abolish this internally antagonistic order. A recognition of the extreme difficulty of this project is inherent in the point on which we opened this chapter, namely, Noys's observation that the military high-technology "endocolonization" of society has been a factor in decomposing the traditional industrial working-class movements that were historically the main agencies of socialist and communist projects. However, there is also a possible reversal of this logic, if averting war, including cyberwar, becomes entwined with other issues, such as struggles for social equality and ecological sustainability, a focal point for recomposition of movements looking beyond capital, drawing on new and diverse constituencies. We have indicated some of the issues that we think might be drawn together around resistance to the rise of cyberwar: antisurveillance sentiment, rejection of the secrecy of the security state and its new digital complexes, concern over the corruption of the general intellect by mis- and disinformation, objection to corporate and military criminality, and, of course, revulsion at the exterminatory horrors of war, from the terror of dirty wars to global thermonuclear catastrophe.

In his reflections on Marxist theories of war and revolution, Balibar (2002) remarks on the coexistence within this body of thought of two contradictory elements, one stressing the idea of "revolutionary war," the other of "revolutionary peace." The first stresses "armed struggle against capital," the second "the refusal of capitalist wars"—"in many respects this class war is therefore also a non-war, or an anti-war" (11). If, in the title to this chapter, we invoked Lenin's *What Is to Be Done?*, it is partly to remember that, though Lenin is today primarily thought of as a theorist of revolutionary war, perhaps the major decision of Leninism was one for peace (even if his later adoption of Trotsky's formula of "no war, no peace" resonates with today's state of cyberwar). The outbreak of World War I precipitated a schism in the international socialist movement. Leading European socialist democratic parties all too rapidly discarded their long-held view that war exemplified the irrationality of competitive capitalism, forgot their commitments to peace and worldwide worker solidarity, rallied behind their governments, and joined the march to mass slaughter. Only the faction of what is sometimes known as the Zimmerwald Left (Nation 1989), led by Lenin, continued to speak out for internationalism. It was thus not just revolt against exploitation but rejection of the holocaust of World War I, a program of "bread and peace," that gave communism a moral claim to universality. Terrifying contemporary parallels to the pre-1914 years today demand an updated strategy for "bread and peace," with "bread" understood as a securing of ecological conditions for species life and "peace" as elimination of systemic social violence.

One of the slogans of the Zimmerwald Left was "Krieg dem Krieg," "war on war," and it is tempting to take this up, in a very literal sense, and propose a "cyberwar on cyberwar." As we have seen, there is a hacking front to both the struggles against digital militarism and contemporary anticapitalist movements. They have exposed the workings of the cyberwar complex and brought it to light. Their main figures are defectors from that complex. More broadly, over recent decades, many critical theorists have argued for forms of "cyborg" dissent and for deployment of the arms available to "immaterial labor," whether in the networked mobilization of protest or in more direct digital disruptions of war making. Cyberwar on cyberwar is both a metaphorical and practical possibility, and we have seen situations when "Krieg dem Krieg," in the most concrete sense, is the only effective response to murderous attack.

However, we would suggest that to conduct "cyberwar on cyberwar," though it may sometimes be necessary, is to fight on unfavorable terrain. We have seen that hacktivism suffers problems of accountability, transparency, provocation; can itself be compromised and ensnared within the exploits of the military–internet complex; and is ultimately highly vulnerable to the police and intelligence apparatus. And, as Noys (2013) observes, even reliance on the speed with which networks can circulate struggles tends to discount how much more advantage such velocity today gives capital's military–security complexes (in this respect, we note that the huge, worldwide, and digitally mobilized protests against the Iraq War in 2003 must be reckoned a tragic failure of networked activism). Cyberwar on cyberwar is a method of fast politics, and speed is where the user-subjects ultimately lose, encountering the inhuman acceleration of machinic processing power. To succeed, such resistance requires a break from the mainstream paradigm of today's "platform capitalism" with all its repetitious "Twitter revolutions," "Facebook revolutions," or "Snapchat revolutions," slogans that should remind us that, as Lacan once notoriously noted, an ultimate misconception of revolution is as a desire for a new master or a master in a new form that leads away from the systemic change rather than not toward it. Here we disagree with the conclusion of Brian Massumi's (2015, 243) otherwise excellent study of the new "ontopowers" of military networking, where, discussing the logic of speed and preemption, he suggests that countermovements have no choice but to "go forward, with the flow."

While tactical resistance can involve any and all of the "memes of production" (Deterritorial Support Group 2012), a reconstitution of the left today must ask, what is the opposite of cyberwar? To this, we would answer that the antithesis of cyberwar is *corporeal care* of the subject achieved through the "balanced conceptions of space and time within culture" and "awareness of spatial and temporal dynamics [that] keep state and market power in check" (Sharma 2013, 314). It is from this perspective that we need to recognize cyberwar's production of time and space and envision different times and spaces—those of the care of bodies. This orientation against the social destruction, physical, psychological, and infrastructural, of cyberwar does not mean totally abandoning the digital—which, because it so much composes the very texture of everyday life, would be not only difficult but often politically fatal. But it does mean its rearticulation to a set of purposes radically different from those of digital capital. In particular, this strategy requires theoretical reconsideration and practical subversion of the addicted, complicit digital user, the figure envisioned by neoliberal Silicon Valley, by way of desynchronization and emancipation.

This can be described as recognizing a position in and against the military environment of cyberwar in which all of us are now imbricated and finding ways to develop subjectivities that are simultaneously of the network and off the network. It requires the "slow" time necessary for the in-person (rather than online) organization of antiwar collectives, movements, and alliances; defection from compulsive social media use; trammeling corporate capacities to intensify and maintain such addictive behavior; the patient defense and reconstruction of the basic public institutions of corporeal care—free health services; the cultivation of mental health; the recovery and deepening of the legacy of a semidestroyed (or, in many places, never created) welfare state in a new "commonfare"; universal education provisions; worker–community control of workplaces and the means of production; ecological protections—and the assertion of such priorities against the expense and logic of networked militarization. In this work of solidarity, the subject exploited and excommunicated by digital capitalism can transition from alienation toward reciprocity. And to those who say that the accelerated logic of cyberwar means we don't have time to do all this before catastrophe arrives, we just say, you may be right, but still we have to do it anyway! We can build a "counterwar machine" constructed on the diagonal line that runs between waging cyberwar on cyberwar and fostering the caring corporeality that is opposite of cyberwar.

At the end of her study of world labor activism, Beverly Silver (2003, 176) notes a major reason for the shortage of militant working-class movements in the early twenty-first century. Neoliberalism's restructuring, globalization, and financialization, with its "growing structural unemployment, escalating inequalities and major disruptions," has repeated the crisis patterns of previous eras of capitalism, with one crucial exception. The missing condition is large-scale armed conflict. This "global political–military context contrasts sharply with . . . that [which] produced radicalized and explosive labour unrest in the first half of the twentieth century." As Silver notes, war then involved the mass mobilization of populations that characterized total war. States depended on their working classes to provide not just millions of soldiers but labor in munition plants, shipyards and aircraft factories, hospitals, and farms. When mass mobilization met the horror of mass deaths and mutilation, revolutionary social turmoil could result.

As Silver (2003, 175) observes, advanced capitalism's turn to high technological weaponry apparently breaks this link between war and worker revolt. Cyberwar can be seen as an extension of this "automation of war." Nonetheless, as we have suggested, the tendency of digital militarization to liquidate the labor of war is not yet completely fulfilled. Humans remain as the indispensable conscious links and relays within the networks and nodes of digital conflict. Indeed, what we have seen in this book is the surprisingly wide diffusion of participation in cyberwarfare, from the highly specialized military and intelligence units at the cutting edge of advanced cyberoperations to strata of mercenary and criminal proxies, online vigilantes, patriotic hackers, corporate and criminal marketers of cyberweaponry, cybersecurity personnel, and on to the corporate content moderators and state censors and surveillance agents now indispensable to the prosecution of war waged in cyberspace and across scores of hybrid battlefields. To these more or less intentional contributions to the mechanisms of cyberwar must be added the unknowing (or partially unknowing) participation of network users, whose online activities and addictions provide the vital vectors for the memes, exploits, and hijackings of subterranean cyberconflicts and whose reconstitution as data-subjects habituated to ceaseless state and commercial surveillance constitutes the inevitable accompaniment to such operations.

Surveying this field, we can say that military mobilization has not so much been abolished from cyberwar as reconfigured in subterranean, etiolated, and unfamiliar forms. This decomposition of the labor of war, equivalent to Virilio's state of "endocolonization" by the apparatus of high-technology militarism, may, as we have proposed in this chapter, contain potentials for reversal. If, to date, cyberwar is not, at least in the centers of capitalism, producing the massive havoc of earlier forms of war, the migrant refugees of hybrid conflicts around the world, fleeing algorithmically directed drones, social media–activated death squads, and cybernetic strikes at social utilities, bear witness to its potential to do so. Already, even in ostensibly secure zones of the planet, the costs of militarized and criminalized networks, in terms of escalating social paranoias, crumbling confidence in everyday communication and polarizing social relations, becomes daily more apparent. If this course persists, unforeseen forms of unrest by the new workforces of cyberwar may interrupt its inhuman trajectory.

ENVOI

As both of us are academics, we are concerned with how the neoliberal university is occupied today by cyberwar. Responding to the exhortations of states and corporations for a supply of labor power adequate to cyberwar conditions, institutions of postsecondary education are proliferating cybersecurity programs (Talley 2013; Ritchie 2016; Wilson 2017). They are also the target of intensifying cyberattacks, some aimed at military-related research (Ismail 2017; Young and Bennett 2017). But, most immediately from our point of view, the university has become a place where students receive mutually contradictory messages from faculty teaching critical theory about the risks of state and corporate surveillance and the public relations teams hired by higher administrators that encourage them to "like" the university's profile on Facebook, follow it on Twitter, and become full participants in a regime of corporate promotion and self-branding. Not only does this jeopardize pedagogical work (except when one manages to use such instances as case studies for politicoeconomic analysis in class, which is admittedly rare) but it is also unethical given the general awareness (including by higher administrators and public relations teams) that youths are being aggressively targeted by corporate platforms. This targeting does not just leave young people feeling stressed, defeated, overwhelmed, anxious, nervous, stupid, silly, useless, and like a failure. It also prepares them as the unthinking data-subject cannon fodder for wars already being waged with computational propaganda, botnets, and the viral relay of virtual weaponry with real material consequence.

We and our students are subjects of the "capitalist unconscious," "the alienated subject *at work* in every discursive action" (Tomšič 2015, 54). Lacan traced the notion of the subject to the beginning of modern science, which initiated the emancipation of the human from theocratic social orders, replacing a ritualistic relation to the world of nature.

Unfortunately, as Samo Tomšič argues, "the emancipatory political potential of scientific revolution" was captured and "neutralized" by a counter-revolutionary capitalism that "needs to be thought of as the restoration of pre-modernity within modernity" (235). This neutralization he characterizes as the construction of "a closed world, marked by totality, finitude and centralization"—the world market, mobilizing atomized, narcissistic, and competitive individuals in never-ending and all-subsuming commodity exchange, a world whose purported eternity and completion negate the perception of "contingency, infinity and instability" that is the true core of scientific emancipation. Pointing to a parallelism between Marx and Lacan, Tomšič observes that capitalist modernity "ceases at the critical point of the subject" (235). He explains,

> While capitalism considers the subject to be nothing more than a narcissistic animal, Marxism and psychoanalysis reveal that the subject of revolutionary politics is an alienated animal, which, in its most intimate interior, includes its other. This inclusion is the main feature of a non-narcissistic love and consequently of a social link that is not rooted in self-love. (233)

The atavistic "pre-modernity within modernity" of capitalism, and the disaster of its capture of advanced science, is nowhere more clearly demonstrated today than by its tendencies toward cyberwar. As the young Althusser (1946, 14) wrote in the midst of the "apocalyptic panic" following the explosion of the first atomic weapons, "the world in which humanity trembles before what it has itself wrought is an extravagant image of the proletarian condition, in which the worker is enslaved by his own labour: it is quite simply, the same world."

Acknowledgments

The authors thank the Rogers Chair in Journalism and New Information Technologies at the Faculty of Information and Media Studies, University of Western Ontario, and Canada's Social Sciences and Humanities Research Council for funding crucial to this project. We thank all the participants in the lecture and seminar series "Information Wars and Information Struggles," which we organized at the University of Western Ontario from 2014 to 2016: Ukrainian political science scholar Volodymyr Kulyk from the Institute of Political and Ethnic Studies, Ukrainian Academy of Sciences; Canadian cybersecurity expert Rafal Rohozinski, founder and CEO of the Sec-Dev Group; war researcher Antoine Bousquet from Birkbeck, University of London; the director of Citizen Lab Ron Deibert and CL researcher Jakub Dalek for an informative workshop on the Citizen Lab's research methods and projects; Craig Forcese, expert legal critic of Canada's national-security legislation, Bill C-51; investigative journalists Andrei Soldatov and Irina Borogan for their insights about risks and futures of digital mobilization and surveillance in Russia; Ukrainian filmmaker Oleksiy Radynski and film producer Lyubov Knorozok for sharing their work filmed in eastern Ukraine on the brink of the war; and the scholar of Anonymous, Gabriella Coleman, McGill University. The discussions with these experts in and veterans of various cyberwar events were foundational to our project.

We also thank Ron Deibert for the invitation to participate in the Citizen Lab workshop on the cyberdimensions of armed conflict with a focus on Russia–Ukraine on August 15–16, 2016, in Toronto. We are grateful to Georgian attorney, former parliamentary secretary to the president, and deputy minister of defense of Georgia, Anna Dolidze, for facilitating interviews with governmental officials and cybersecurity specialists about the events of the 2008 Russo-Georgia cyberwar in Tbilisi in September 2016; and to David Lee, president of MagtiCom, a major Georgian tele-communication and internet company, for sharing his insights on the changing relation between media and the government in Georgia over several decades. We thank the members and associates of the Visual Culture Research Institute, in particular Oleksiy Radynski, who hosted us so generously in Kyiv in October 2015, discussed Ukraine's experiences of cyberwar, and helped us arrange a series of interviews with participants in that history that were profoundly important to our thinking on the topic. The conversations about the ongoing cyberattacks on the Ukrainian internet and physical infrastructure with Oleksiy Misnik, a government security expert at the Ukrainian Cybercenter and CERT-UA, in February and November 2017 in Kyiv, were very insightful. We very much appreciated Voldodymr Ishchenko's invitation to speak at the National Technical University of Ukraine and the conversations that followed.

We thank the conference and speaker series organizers who invited us to present on the topic of our book: Nandita Biswas-Mellamphy (Electro-Governance Research Group workshop in January 2016); Adam Kingsmith, Robert Latham, and Julian von Bargen (*Augmenting the Left* in May 2017); and Daphne Dragona, Jussi Parikka, and Ryan Bishop (*transmediale* in February 2018).

Finally, we thank Danielle Kasprzak at the University of Minnesota Press for so swiftly and strongly supporting the publication of our work.

Nick Dyer-Witheford thanks Svitlana Matviyenko for good years of friendship and of thinking, researching, and writing together. He thanks Thomas Carmichael, James Compton, Atle Kjøsen, Victoria Rubin, and Gil Warren for their interest in, and suggestions for, this project. He especially thanks his wife, Anne Dyer-Witheford, for her loving care and her support of the research and writing of this book.

Svitlana Matviyenko is grateful to Nick Dyer-Witheford, a friend and coauthor of this book, for taking the lead in assembling the manuscript and being an inspirational collaborator throughout several years of educating and enriching work on this project.

Notes

1 The only in-depth study of the issue is in Wikiquote (https://en.wikiquote
   .org/wiki/Leon_Trotsky). It concludes that the "quotation" is at best a
   loose interpretation of a purported retort by Trotsky to James Burnham:
   "you may not be interested in dialectics, but the dialectic is interested in
   you." The issue is further muddied because, while Wikipedia identifies
   novelist Alan Furst as responsible for the attribution, it mistakes the novel
   in which it occurs as his *Night Soldiers* (1988) rather than *Dark Star* (1991),
   where it clearly appears on the opening page.
2 Among these, Bruce Schneier's (2018) *Click Here to Kill Everybody: Security
   and Survival in a Hyper-connected World* is authoritative on the technical
   aspects of cybersecurity.
3 According to the *OED*, this term originated in early-1950s America in reference to communist sympathizers influenced by Soviet propaganda. It is
   often attributed to Vladimir Lenin, but, once again, there is no evidence
   he authored the phrase.
4 In mentioning these cases together, we are aware that a proxy war localized in the Ukrainian east is different from Syrian "total" proxy war of all
   against all, although the possibility of the development of a war between
   the Russian Federation and the United States held on territory in Ukraine
   was rather high in 2014.

1. THE GEOPOLITICAL AND CLASS RELATIONS OF CYBERWAR

1 An excellent overview of Marxist thought on war is in Egan (2013, 1–10).
   Draper and Haberkern (2005) provide a collection of and commentary on
   Marx and Engels's writings on war.
2 This account of Cold War cyberwar leaves aside one controversial topic: the
   suggestion that in the early 1980s, the CIA successfully implanted a "logic

bomb" in the software of a Soviet gas pipeline's control system and caused a malfunction that resulted in a monumental explosion visible from space. If this occurred, it would be the first critical infrastructure-disabling cyber-attack on record, predating Stuxnet by some thirty years. Most Western cyberwar experts have dismissed reports of the alleged event, but Klimburg (2017, 135), reassessing the event, is "inclined to think something like it did occur."

3   For a discussion of Marxism's theories of international relations, see Teschke (2008).

4   For example, from the 1980s on, an impressive number of teenage hackers have penetrated U.S. military systems, often causing great alarm. Some instances include the 1990 break-in by young Dutch hackers into three dozen military computers, seeking information on U.S. Patriot missiles; the 1994 breach by a sixteen-year-old, who entered both U.S. Air Force networks and those of the South Korean nuclear research center; and the event dubbed by U.S. security agencies Operation Solar Sunrise, in which, in the midst of the Iraq War, two Californian teenagers, assisted by an Israeli hacker, broke into some five hundred military systems (Zetter 2014a). In all cases, these intrusions were initially believed to be the work of hostile states.

5   On the long debate over the Revolution in Military Affairs, see Gongora and von Riekhoff (2000), Knox and Murray (2001), Sloan (2002), Gray (2004), Metz and Kievit (2013), and U.S. Army Command and General Staff College (2014).

6   In a detailed analysis of Snowden-released documents, Ryan Gallagher and Glen Greenwald (2014) suggest that the NSA has possessed capacities to implant malware in computers since 2004 but operated on a relatively small scale. By 2012, however, its TAO unit was tasked with "aggressively scal[ing]" these operations, so that the number of implants rose from the low hundreds to tens of thousands. Even this enlargement was, however, insufficient; the NSA determined that "managing a massive network of implants is too big a job for humans alone," because human "drivers'" tendency to "operate within their own environment, not taking into account the bigger picture," limited their "ability for large-scale exploitation." To overcome this barrier, the NSA created the TURBINE system, an "intelligent command and control capability" for "industrial-scale exploitation," designed to "allow the current implant network to scale to large size (millions of implants) by creating a system that does automated control implants by groups instead of individually."

7   In 2018, the *Wall Street Journal* reported that the Trump administration planned loosening policy restraints on U.S. cyberattacks (Volz 2018).

8   For discussion of the outburst of opposition by workers at these companies to such military and paramilitary involvements, see chapter 3.

9   In 2018, a report from the whistle-blowing blog *The Intercept* (R. Gallagher

2018) disclosed that Google was secretly developing a plan to relaunch its China involvement. Code-named "Project Dragonfly," it involves about three hundred staff and entails working in collaboration with a Chinese company and observing China's censorship regulations. News of Dragonfly was greeted with dismay and protest by many Google employees and widely criticized in the United States. What is not yet evident is China's military's and security agencies' view of Google's potential reinsertion into their nation's digital networks.

10 Some 2018 reports claim a resurgence in hacking attacks from China against U.S. targets, at heightened levels of sophistication and destructiveness (S. Gallagher 2018b). An ongoing front in cyberhostilities between the two nations concerns malware or secret vulnerabilities added to computer, mobile phone, and telecommunication equipment exported from one nation to another. As early as 2012, the House of Representatives's intelligence committee had advised American companies and its government to avoid doing business with China's two leading technology firms, Huawei and ZTE, because they pose a national security threat, and said regulators should block mergers and acquisitions in the United States by the two companies. The Trump administration raised this issue to a new level by exhorting U.S. corporations and educational institutions, and also foreign allies, such as Canada, to sever business and research relations with these companies. In 2018, Congress passed a bill banning government use of Huawei and ZTE technologies for essential services (Kastrenakes 2018). Meanwhile, Edward Snowden's disclosures and sources suggest that U.S. suspicions of Huawei and ZTE were partly based on the success of the NSA in infiltrating China's networks by methods similar to those the Chinese corporations were accused of using (Hsu 2014).

11 Gerasamov's article caught the attention of Prague-based war researcher Mark Galeotti, who somewhat recklessly termed it a "doctrine." In the face of subsequent criticism, Galeotti conceded that Gerasamov's article was not a "foundational strategy document" and that it was a "mistaken and dangerous assumption" to believe that "what [Russians] are doing is both new and truly distinctive and, by implication, uniquely threatening" (Galeotti 2018, 2, 3)—even if, by accident, Gerasimov's piece became a useful weapon for both U.S. and Russian information warriors: for the former, raising the specter of a New Cold War, and for the latter, claiming ownership of a strategy in fact partly modeled on Western example.

12 Although Russia and China are both described as threats to U.S. hegemony, the nature of their challenge is different: while China is growing in economic power, the Russian economy is, other than in its energy sector, deteriorating. It is reported that while the U.S. intelligence agencies consider Russian hackers more skilled than Chinese hackers, they regard the latter as more of a long-term threat, because China, unlike Russia, now has a thriving

national digital industry and is home to corporations, such as Alibaba, Tencent, and Baidu, that are among the most important platforms on the planet and have a larger "footprint" from which to conduct cyberoperations (BBC 2018a).

13 We return to the notion of "immaterial labor" in chapter 3, where we focus on the affective and emotional aspects of another type of a worker, the "user," to argue that such labor is material in every way.

14 Among the large body of literature on hacking, see, in addition to sources mentioned in the text, Jordan and Taylor (2004), Jordan (2008), and Coleman (2015). For Marxist arguments that the paradoxes of hacking are characteristic of middle-class "contradictory class positions," see Dyer-Witheford (2015) and Steinmetz (2016).

15 One of these suspected jihadist cyberattacks nearly destroyed a French TV network, V5Monde, which was taken off-air on April 8, 2015, after it was penetrated by malware. A group calling itself the Cyber Caliphate, linked to ISIS, claimed responsibility, but a subsequent investigation suggested that the attack was carried out by the Russian hacker group APT 28, later to become notorious for hacking the U.S. Democratic National Convention as a disguised test run of its cyberwar weaponry (Corera 2015). This exemplifies the attribution problems characteristic of cyberwar.

16 For conflicting assessments of the politics of the Ukraine–Russia conflict, see Wilson (2014), Sakwa (2014), Mennon and Rumer (2015), and Toal (2017).

17 To give just one example, a report by *Bloomberg Businessweek* in 2013 features Andrés Sepúlveda, currently "serving 10 years in prison for charges including use of malicious software, conspiracy to commit crime, violation of personal data, and espionage, related to hacking during Colombia's 2014 presidential election" using $50,000 worth of "high-end Russian software that made quick work of tapping Apple, BlackBerry, and Android phones" and hosting his databases on now-erased "servers in Russia and Ukraine rented anonymously with Bitcoins" (Robertson, Riley, and Willis 2016). The international cybercrime teams assembled by Sepúlveda ("Brazilians, in his view, develop the best malware. Venezuelans and Ecuadoreans are superb at scanning systems and software for vulnerabilities. Argentines are mobile intercept artists. Mexicans are masterly hackers in general but talk too much") were working on assisting right-wing candidates "in presidential elections in Nicaragua, Panama, Honduras, El Salvador, Colombia, Mexico, Costa Rica, Guatemala, and Venezuela" and overall had "enough wins that he might be able to claim as much influence over the political direction of modern Latin America as anyone in the 21st century" (Robertson, Riley, and Willis 2016). In particular, he and his teams "splurged on the very best fake Twitter profiles; they'd been maintained for at least a year, giving them a patina of believability": "Sepúlveda managed thousands of such fake profiles and used the accounts to shape discussion around topics such as Peña

Nieto's plan to end drug violence, priming the social media pump with views that real users would mimic. For less nuanced work, he had a larger army of 30,000 Twitter bots, automatic posters that could create trends. One conversation he started stoked fear that the more López Obrador [a Mexican left-wing politician] rose in the polls, the lower the peso would sink. Sepúlveda knew the currency issue was a major vulnerability; he'd read it in the candidate's own internal staff memos" (Robertson, Riley, and Willis 2016).

18 Throughout 2018, the U.S. government and major technology corporations produced increasingly detailed evidence of Russian interference in the election. In June, the Justice Department (2018b) issued a second indictment against twelve members of the Main Intelligence Directorate of the Russian General Staff (better known as *Glavnoye razvedyvatel'noye upravleniye,* or GRU) claimed to have conducted "hacking offenses" in an election tampering campaign years in the preparation and operating with a substantial budget (S. Gallagher 2018a; Goodin 2018). In October 2018, Twitter made public a "troll tweet trove" of 10 million tweets posted by suspected state-backed Russian and Iranian "troll farms" between 2013 and 2018, reportedly making reference to both the U.S. election and the U.K. referendum on Brexit (BBC 2018g). Meanwhile, the actual effects of such activities continue to be debated, with Kathleen Hall Jamieson's (2018) *Cyberwar: How Russian Hackers and Trolls Helped Elect a President—What We Don't, Can't, and Do Know* making the case that they contributed significantly to Trump's victory.

19 In light of the subsequent Cambridge Analytica scandal, Facebook in 2018 announced it would no longer send employees to work at the offices of political campaigns during elections (BBC 2018e).

2. CYBERWAR'S SUBJECTS

1 This example is from a report by the cybersecurity company FireEye on hacking in the Syrian civil war (Reglado et al. 2015) and is discussed further later.

2 The photographs of the messages on the phone of a Ukrainian army soldier were tweeted by Radio Free Europe/Radio Liberty journalist Christopher Miller, who covered the war in the Donbas region, on February 2, 2017: https://twitter.com/christopherjm/status/827157599325941766.

3 As we have seen, the image of the Mongol steppe warriors was central to Deleuze and Guattari's concept of a nomad "war machine." Arquilla and Ronefeldt's (1993) landmark RAND report "Cyberwar Is Coming!," which in many ways initiated recent Pentagon thinking about war in the networks, opens with the idea of Mongol "hordes" (24). Arquilla and Deleuze cite DeLanda and, in a later document, Deleuze and Guattari. This

historical image of alien threat was very soon combined with an example of netwar waged by a contemporary, but equally foreign, insurgency: that of the Mayan Zapatista uprising, with its online, digital world circulating manifestos of Subcommandate Marcos. Deleuze and Guattari had always recognized that the nomad heterodox war machine could be appropriated or domesticated by the state, making it an aspect or limb of orthodox military power. This is precisely what the RAND thinkers, and their many successors, set out to do. The IDF has conducted similar military appropriation of Deleuze and Guattari; see Weizman (2012).

4 For an important study of "mobilization," see the work of Jean-Paul de Gaudemar (1979) and its application to information capitalism and techno-war by Kevin Robins and Frank Webster.

5 The military conflict in Donbas created the possibilities for lucrative contracts in the conflict zone used mainly by businessmen associated with Ukrainian government. This was permitted in the context of Anti-Terrorist Operations but would have been illegal if the term *war* had been accepted earlier. Because the term *civil war* has been aggressively used by the Russian propaganda channels, in Ukraine, such definition has been almost univocally denied. However, as David Armitage observes, "civil wars have been so paradoxically fertile because there has never been a time when their definition was settled to everyone's satisfaction or when it could be used without question or contention. This is in part because conceptions of civil war have been disputed and debated within so many different historical contexts. . . . This problem of naming becomes particularly acute when political ideas are at stake. . . . When the framing term is one like 'civil war,' however, politics precedes even attempts at definition" (Armitage 2017, "Confronting Civil War"). In our discussion of cyberwar, we want to second Armitage's assertion that "civil war has gradually become the most widespread, the most destructive, and the most characteristic form of organized human violence," and as he quotes the U.S. Civil War general William Tecumseh Sherman, "war is hell . . . , but surely the only thing worse is civil war" (Armitage 2017, "Confronting Civil War"). Civil wars almost never occur without foreign intrusion or assistance. Over the last decades, they have become the messiest outcomes of cyberwars, the evidence that despite their cyberdimension, they possess a material core: "civil wars are like a sickness of the body politic, destroying it from within" (Armitage 2017, "Confronting Civil War"). Therefore to identify a military conflict as "civil war" means acknowledging its messiness, complexity, and criminal responsibility of all proxies.

6 Misattribution of "proofs" has become unprecedented and, in some cases, a new way of trolling. American director Oliver Stone's TV series *The Putin Interviews* features an episode where Putin shows Stone a video that he claims demonstrates "the work of the Russian aviation in Syria." The

footage was immediately identified as in fact a video of American aviators attacking the Taliban in 2012. The original sound track, though, is removed and reportedly substituted with recorded voices of Ukrainian pilots (speaking in Russian) during the military operations in the current Donbas war. The episode indicates the morass of deception and misrecognition made possible by digital production technologies.

7 Journalist Katerina Sergatskova observes that in Donbas, "working without accreditation was a surefire invitation to prison and worse" and asks, "Has Ukraine become more dangerous for journalists than Russia?" She refers to the ongoing surveillance of the journalists of *Ukrainska Pravda*, threatened on multiple occasions by receiving the proofs of surveillance: hacked emails and phone text records (Sergatskova 2016).

8 Althusser's (1971) inadequate answer to this issue is that it doesn't matter if the institutions composing the ISAs are "public" or "private" and that "private institutions can perfectly well 'function' as Ideological State Apparatuses" (144).

9 For interviews with people who experienced this recruitment process, see Patrikarakos (2017, 203–30) and also the film by Norwegian directors Adel Khan Farooq and Ulrik Imtiaz Rolfsen *Recruiting for Jihad* (2017).

10 This, some NATO analysts of Russian trolling suggest, is an elaborately orchestrated team process in which one troll will initiate a topic and another will follow up with an extreme counterview to inflame a situation—but others hold that this is just a series of knock-on effects.

11 In a broad optic, this martial reinterpretation of interpellation should also extend to "class war." Stuart Hall and other early cultural studies theorists attempted to synthesize Althusser's theory of interpellation with the concept of "hegemony" and "counterhegemonic struggles" derived from Gramsci—who, with his metaphors of "wars of position" and wars of maneuver," is one of the most military of Marxist theorists.

12 It is notable that Althusser's famous example of the police "hey you" interpellation actually crosses this line between "ideological" and "repressive/violent" state apparatuses. As Black Lives Matter reminds us, for some subjects, failure, or even success, in properly answering this summons is likely to be met with a volley of bullets.

13 Writing in the context of the #MeToo movement, Ava Kofman (2018) argues than an important element in this exclusion is a "toxic culture of sexism and harassment at cybersecurity conferences." Sexist bias seems in play even in the underreporting of female cyberwar and cybersecurity whistleblowers like Reality Winner (Maas 2018).

14 The study, from the security firm Incapsual, was based on data collected from one thousand websites that utilized the firm's services; it determined that "just 49% of Web traffic is human browsing. 20% is benign non-human search engine traffic, but 31% of all Internet traffic is tied to malicious

activities. 19% is from 'spies' collecting competitive intelligence, 5% is from automated hacking tools seeking out vulnerabilities, 5% is from scrapers and 2% is from content spammers" (van Mensvoort 2012).

15   A similar trajectory seems to have been followed by the Flame reconnaissance and espionage virus, likely developed by the same team that created Stuxnet, which infected computers in Iran, the West Bank, Sudan, Syria, Lebanon, Saudi Arabia, and Egypt.

16   On May 29, 2018, the world's networked news outlets reported that Russian journalist Arkady Babchenko had been gunned down and found bleeding at the entrance of his apartment in Ukraine's capital, Kyiv. It was suggested he had been targeted because of his fierce criticism of the Kremlin for the annexation of Crimea and support of separatists in southeast Ukraine. On the next day, however, the head of the Ukrainian Security Service (SBU), Vasyl Hrytsak, revealed that Babchenko was alive and the whole episode, including the photographs of the "dead" journalist, had been staged. According to Hrytsak, the event was masterminded by the SBU to prevent and expose an actual Moscow-planned assassination attempt on Babchenko by prompting the assassin to contact his employers, spuriously claim success, and receive a payment. Millions of human network users—the whole of the planetary Stack (Bratton 2016), one might say—were used as relays to deliver the fake event and enable the SBU to fish for a subject of interest by sacrificing the credibility of digital news media, a technique adopted from the northern neighbor it constantly criticizes (Harding and Roth 2018).

17   *Le nom-du-père,* to use Lacan's term for the law of the Symbolic order, which he rephrased, playing on the homonymic similarities, as *le non du père,* or "the no of the father."

18   In *Seminar III,* Lacan ([1967] 1997) defines a signifier as that which represents a subject for another signifier by opposing it to the sign, which represents something for someone.

3. WHAT IS TO BE DONE?

1   The concept of class composition is associated with the schools of workerism *(operaismo)* and autonomist Marxism. It derives from Marx's concept of the organic composition of capital—the ratio of fixed capital (raw materials and machines) to variable capital (workers). In Marx's account, this ratio was considered as it affected the waxing and waning of capital's surplus value extraction. *Operaismo* famously "inverted" this idea to view it from the position of the working class. The "technical composition" of the class involved the modes of division, control, and replacement imposed on it by capital, but its "political composition" was the organizational capacity of workers to resist capital. The role of technology is crucial, not in any determinist sense, but understood as the "weaponry" of class war wielded by capital

to "decompose" or disintegrate the political composition of the working class. Autonomists suggest that capitalist attacks on and decompositions of the working class are cyclically followed by working-class recompositions. Noys (2011), however, is drawing on a bleaker version of class-composition theory enunciated by recent "communization" tendencies, such as Theorie Communiste and EndNotes, in whose view the decomposition of the industrial working class in the late twentieth century has not been followed by a recompositionary moment.

2 It is a sign of surveillant times that the online news site The Intercept, created in the aftermath of Snowden's leaks by his collaborator, Glen Greenwald, has become an important organ of Trump-era dissent in the United States.

3 Anonymous activists have been especially hard hit. Some twenty-five members involved in Operation Avenge Assange were arrested in the United States, the United Kingdom, and the Netherlands, largely on the basis of evidence from a prominent Anon turned FBI informant. The same year, some thirty-two Anons were arrested in Turkey for alleged involvement in DDoS attacks on Turkish government websites (Albanesius 2011). The Free Anons campaign of the Anonymous Solidarity Network continues to offer support to Anons arrested around the world. In 2013, a British Finnish hacker, Lauri Love, was arrested for breaching the security of U.S. government sites, including those of the U.S. Army, Missile Defense Agency, and NASA, extracting sensitive information about tens of thousands of employees (BBC 2014). The attack was accompanied by a video reportedly "declaring war" on the U.S. judicial system for its treatment of hackers. At the time of writing, a U.K. high court has just granted Love's appeal against a U.S. extradition request, which he contested on the grounds of "unconscionable" U.S. prison conditions.

4 As Gray (2017) explains, "operational art," or in Russian, оперативное искусство, is a level of war between battlefield tactics and grand strategy. It deals with the coordination of activities at the level of a specific front or theater of war. Although operational art has been used for centuries, it was marshal of the Soviet Union (MSU) Mikhail Nikolayevich Tukhachevsky (1893–1937) who codified it as a systemic concept in his Provisional Field Regulations of 1936 for the Soviet army. The marshal was killed in the Stalinist purges, and his theories were abandoned. However, "recent developments, including . . . the Revolution in Military Affairs have given [them] a new and permanent lease on life" (Gray 2017).

5 The story is additionally complicated by allegations that the Shadow Brokers, working for the Russian government, stole the files after they had been identified on the home computer of a NSA employee through the use of antivirus software from Moscow-based cybersecurity firm Kaspersky Lab (Goodin 2017).

6 On a more theoretical level, Nunes challenges the orthodox interpretation

of Deleuze and Guattari's work as positing a dichotomy between "rhizom-atic" (horizontal) and "arborescent" (vertical) formations and suggests that their discussion of the lupine "pack" actually posits an intermediate type of "distributed leadership." In this respect, Michael Hardt and Antonio Negri's revocation of their entirely "spontaneous" concept of revolution (Hardt and Negri 2000) and reconsideration of the role of "the Prince" (Gramsci's code for the party) (Hardt and Negri 2009), and their more emphatic assertion of a need for movement leadership, but at a tactical level rather than strategic level (Hardt and Negri 2017), are significant. Also important in this reconsideration of the role of networks in movements is Geert Lovink and Ned Rossiter's concept "org.nets" (Rossiter 2006).

7 We make no bets on the outcome of this historical moment. Some observers see the United States in irreversible decline; others insist on its continuing ascendancy. Alfred McCoy (2017), who argues for the erosion of U.S. pow-ers, suggests five scenarios: a subsidence of U.S. hegemony into a more multipolar world; tumultuous economic decline; eventual military defeat in overextended foreign interventions and hybrid conflicts; full-scale war with China; and disintegration in the midst of a global warming crisis. Only the first of these could be considered a "soft landing'"; all the others involve dire military situations in which cyberwar is a component, prelude, and, quite possibly, consummation.

8 Elaine Scarry (2014), in her critique of what she terms *thermonuclear monarchy,* highlights the situation of the United States's thirteen Ohio class nuclear submarines, each carrying the equivalent of "4000 Hiroshima bombs" (367). These submarines, "in wartime or time of great political tension," remain submerged at a depth where they can only receive commands via extremely low frequency (ELF) waves that can penetrate the ocean depths to deliver a very simple and slow "bell ringer" message summoning the submarine closer to the surface, where it can receive, via an airborne relay, more complex messages. It is believed, though not confirmed, that ELF could also be the relay for actual launch commands. However, Scarry notes, "so possible is it that even this message will not get through" that standing navy procedures delegate launch authority over nuclear weapons to ship commanders "without an order from the civilian government" (11–12). The possibility of ELF transmissions being sabotaged in cyberwar conflicts has recently been raised (Thomas-Noone 2017), highlighting the fragility of nuclear command, control, and communication systems.

Bibliography

Abbate, Janet. 1999. *Inventing the Internet*. Cambridge, Mass.: MIT Press.

Abbate, Janet. 2003. "Women and Gender in the History of Computing." *IEEE Annals of the History of Computing*. https://pdfs.semanticscholar.org/e7cc/e3470da4f949806612f6ab105f3006485a8c.pdf.

Abbate, Janet. 2013. *Recoding Gender: Women's Changing Participation in Computing*. Cambridge, Mass.: MIT Press.

Ackerman, Spenser. 2014. "Snowden: NSA Accidentally Caused Syria's Internet Blackout in 2012." *Guardian*, August 13. http://www.theguardian.com/world/2014/aug/13/snowden-nsa-syria-internet-outage-civil-war.

Ackerman, Spencer. 2015. "Barrack Obama and Surveillance Reform: A Story of Vacillation, Caution and Fear." *Guardian*, June 3. https://www.theguardian.com/us-news/2015/jun/03/barack-obama-surveillance-reform-vacillation-caution-fear.

Åhäll, Linda. 2015. *Sexing War/Policing Gender*. New York: Routledge.

Albanesius, Chloe. 2011. "Turkey Arrests 32 'Anonymous' Members & Opinion." *PCMag*, June 13. https://www.pcmag.com/article2/0,2817,2386803,00.asp.

Albright, Jonathan. 2016a. "The #Election2016 Micro-Propaganda Machine." *Medium,* November 18. https://medium.com/@d1gi/the-election2016-micro-propaganda-machine-383449cc1fba/.

Albright, Jonathan. 2016b. "How Trump's Campaign Used the New Data-Industrial Complex to Win the Election." *LSE US Centre* (blog), November 26. http://blogs.lse.ac.uk/usappblog/2016/11/26/how-trumps-campaign-used-the-new-data-industrial-complex-to-win-the-election/.

Aldridge, Robert. 1999. *First Strike! The Pentagon's Strategy for Nuclear War*. New York: South End.

Alexander, David. 2015. "Pentagon Teams Up with Apple, Boeing to Develop Wearable Tech." *Reuters Business News,* August 28. https://www.reuters.com/article/us-usa-defense-tech/pentagon-teams-up-with-apple-boeing-to-develop-wearable-tech-idUSKCN0QX12D20150828/.

Aliaksandrau, Andrei. 2014. "Brave New War: The Information War between Russia and Ukraine." *Index on Censorship* 43: 54–60.

Alliez, Éric, and Maurizio Lazzarato. 2018. *Wars and Capital.* New York: Semiotext(e).

Althusser, Louis. 1946. "The International of Decent Feelings." In *The Spectre of Hegel: Early Writings,* edited by François Matherson, 21–23. London: Verso.

Althusser, Louis. 1971. *Lenin and Philosophy and Other Essays.* London: Verso.

Amin, Samir. 2010. *The Law of Worldwide Value.* New York: Monthly Review.

Amin, Samir. 2013. "China 2013." *Monthly Review* 64, no. 10. https://monthly review.org/2013/03/01/china-2013/.

Amoore, Louise. 2009. "Algorithmic War: Everyday Geographies of the War on Terror." *Antipode* 41, no. 1: 49–69.

Amoore, Louise. 2013. *The Politics of Possibility: Risk and Security beyond Probability.* Durham, N.C.: Duke University Press.

#antisec. 2016. "Hack Back: A DIY Guide." http://pastebin.com/raw/0SNSvyjJ.

Aouragh, Miryam. 2003. "Cyber Intifada and Palestinian Identity." *ISIM Newsletter* 12. https://openaccess.leidenuniv.nl/bitstream/handle/1887/16852/ISIM_12_Cyber_Intifada_and_Palestinian_Identity.pdf.

Apprich, Clemens. 2017. *Technotopia: A Media Genealogy of Net Cultures.* Kindle ed. Lanham, Md.: Rowman and Littlefield International.

Apuzzo, Matt, and Sharon LaFraniere. 2018. "13 Russians Indicted as Mueller Reveals Effort to Aid Trump Campaign." *New York Times,* February 16. https://www.nytimes.com/2018/02/16/us/politics/russians-indicted-mueller-election-interference.html/.

Arkin, William, Ken Dilanian, and Cynthia McFadden. 2016. "What Obama Said to Putin on the Red Phone about the Election Hack." *NBC News,* December 19. http://www.nbcnews.com/news/us-news/what-obama-said-putin-red-phone-about-election-hack-n697116.

Armitage, David. 2017. *Civil Wars: A History in Ideas.* Toronto, Ont.: Penguin Canada.

Arquilla, John. 2012. "Cyberwar Is Already upon Us: But Can It Be Controlled?" *Foreign Policy,* February 27. http://foreignpolicy.com/2012/02/27/cyberwar-is-already-upon-us/.

Arquilla, John, and David Ronfeldt. 1993. "Cyberwar Is Coming!" *Comparative Strategy* 12, no. 2: 141–65.

Arquilla, John, and David Ronfeldt. 1996. *The Advent of Netwar.* Santa Monica, Calif.: RAND.

Arquilla, John, and David Ronfeldt. 1997. *In Athena's Camp: Preparing for Conflict in the Information Age.* Santa Monica, Calif.: RAND.

Arquilla, John, and David Ronfeldt. 2000. *Swarming and the Future of Conflict.* Santa Monica, Calif.: RAND.

Arrighi, Giovanni. 2010. *The Long Twentieth Century: Money, Power and the Origins of Our Times.* London: Verso.

Ashby, W. Ross. (1963) 1966. *An Introduction to Cybernetics.* New York: John Wiley.

Assange, Julian. 2012. *CypherPunks: Freedom and the Future of the Internet.* New York: O/R Books.

Assange, Julian. 2014. *When Google Met Wikileaks.* New York: OR Books.

Atwan, Ban Ardel. 2015. *Islamic State: The Digital Caliphate.* London: Saqui Press.

Balibar, Étienne. 2002. "Marxism and War." *Radical Philosophy* 160: 9–18.

Balibar, Étienne. 2015. "In War." *Open Democracy,* November 16. https://www.opendemocracy.net/can-europe-make-it/etienne-balibar/in-war/.

Barbrook, Richard, and Andy Cameron. 1996. "The Californian Ideology." *Science as Culture* 26: 44–72.

Barlow, John Perry. 1996. "A Declaration of the Independence of Cyberspace." Electronic Frontier Foundation. February 8. https://www.eff.org/cyberspace-independence/.

Bartles, Charles. 2016. "Getting Gerasimov Right." *Military Review,* January–February. https://community.apan.org/cfs-file/__key/docpreview-s/00-00-00-11-18/20151229-Bartles-_2D00_-Getting-Gerasimov-Right.pdf.

Barton, Ethan. 2016. "House Watchdog Says Cyber-security War Can't Be Won with Men Only." *Daily Caller,* August 10. http://dailycaller.com/2015/10/08/house-watchdog-says-cyber-security-war-cant-be-won-with-white-men-only/.

Bauman, Zygmunt, Didier Bigo, Paulo Esteves, Elspeth Guild, Vivienne Jabri, David Lyon, and R. B. J. Walker. 2014. "After Snowden: Rethinking the Impact of Surveillance." *International Political Sociology* 8: 121–44.

BBC. 2014. "Briton Lauri Love Faces New US Hacking Charges." *BBC News,* February 27. http://www.bbc.com/news/world-us-canada-26376865.

BBC. 2018a. "CIA Chief Says China 'as Big a Threat to US' as Russia." *BBC News,* January 30. http://www.bbc.com/news/world-us-canada-42867076.

BBC. 2018b. "Facebook Facial Recognition Faces Class-Action Suit." *BBC News,* April 17. http://www.bbc.com/news/technology-43792125.

BBC. 2018c "Facebook to Exclude Billions from European Privacy Laws." *BBC News,* April 19. http://www.bbc.com/news/technology-43822184.

BBC. 2018d. "Russia–Trump Inquiry: Russian Foreign Minister Dismisses FBI Charges." *BBC News,* February 17. http://www.bbc.com/news/world-us-canada-43095881/.

BBC. 2018e. "Facebook Stops Sending Staff to Help Political Campaigns." *BBC News,* September 21. https://www.bbc.com/news/technology-45599962.

BBC. 2018f. "Google Drops $10bn Battle for Pentagon Data Contract." *BBC News,* October 9. https://www.bbc.com/news/technology-45798153.

BBC. 2018g. "Twitter's 'Russia-Iran' Troll Tweet Trove Made Public." *BBC News,* October 17. https://www.bbc.com/news/world.

Benjamin, Medea. 2103. *Drone Warfare: Killing by Remote Control.* New York: Verso.

Benkler, Yochai, Robert Faris, Hal Roberts, and Ethan Zuckerman. 2017.

"Breitbart-Led Right-Wing Media Ecosystem Altered Broader Media Agenda." *Columbia Journalism Review,* March 3. https://www.cjr.org/analysis/breitbart-media-trump-harvard-study.php.

Berardi, Franco "Bifo." 2016. "The Coming Global Civil War: Is There Any Way Out?" *e-flux* 69. https://www.e-flux.com/journal/69/60582/the-coming-global-civil-war-is-there-any-way-out/.

Berger, J. M. 2015a. "The Metronome of Apocalyptic Time: Social Media as Carrier Wave for Millenarian Contagion." *Perspectives on Terrorism* 9, no. 4. http://www.terrorismanalysts.com/pt/index.php/pot/article/view/444/html.

Berger, J. M. 2105b. "Tailored Online Interventions: The Islamic State's Recruitment Strategy." Combating Terrorism Centre. October 23. https://ctc.usma.edu/posts/tailored-online-interventions-the-islamic-states-recruitment-strategy/.

Bertaud, Jean-Paul. 1988. *The Army of the French Revolution: From Citizen Soldiers to Instruments of Power.* Princeton, N.J.: Princeton University Press.

Biddle, Sam. 2017. "Trump's 'Extreme-Vetting' Software Will Discriminate against Immigrants 'under a Veneer of Objectivity,' Say Experts." *The Intercept,* November 16. https://theintercept.com/2017/11/16/trumps-extreme-vetting-software-will-discriminate-against-immigrants-under-a-veneer-of-objectivity-say-experts/.

Bieler, Andreas, Werner Bonefeld, Peter Burnham, and Adam David Morton. 2006. *Global Restructuring, State, Capital and Labour: Contesting Neo-Gramscian Perspectives.* London: Palgrave.

Bieler, Andreas, and Adam David Morton. 2018. *Global Capitalism, Global War, Global Crisis.* Cambridge: Cambridge University Press.

Bienaimé, Pierre. 2014. "This Chart Shows How the US Military Is Responsible for Almost All the Technology in Your iPhone." *Business Insider,* October 29. http://www.businessinsider.com/the-us-military-is-responsible-for-almost-all-the-technology-in-your-iphone-2014-10.

Black, Jeremy. 2006. *The Age of Total War, 1860–1945.* New York: Praeger.

Blue, Violet. 2017. *How to Be a Digital Revolutionary.* Kindle ed. Digital Publications.

Blunden, Bill, and Violet Cheung. 2014. *Behold a Pale Farce: Cyberwar, Threat Inflation and the Military Industrial Complex.* Walterville, Oreg.: TrineDay.

Bolsover, Gillian, and Phillip Howard. 2017. "Computational Propaganda and Political Big Data: Moving toward a More Critical Research Agenda." *Big Data* 5, no. 4: 273–76. https://doi.org/10.1089/big.2017.29024.cpr.

Boucher, Geoff. 2017. "The Long Shadow of Leninist Politics: Radical Strategy and Revolutionary Warfare after a Century." *Philosophical Studies in Contemporary Culture* 25: 141–59.

Bousquet, Antoine. 2008. "Cyberneticizing the American War Machine: Science and Computers in the Cold War." *Cold War History* 8, no. 1: 77–102.

Bousquet, Antoine. 2009. *The Scientific Way of Warfare.* New York: Columbia.

Bousquet, Antoine. 2012. "Complexity Theory and the War on Terror: Understanding the Self-Organising Dynamics of Leaderless Jihad." *Journal of International Relations and Development* 15: 345–69.

Bowden, Mark. 2011. *Worm: The First Digital World War.* New York: Atlantic.

Bradshaw, Samantha, and Philip N. Howard. 2017. "Troops, Trolls and Troublemakers: A Global Inventory of Organized Social Media Manipulation." Working Paper 2017.12, Computational Propaganda Research Project, Oxford University.

Bratton, Benjamin. 2016. *The Stack: On Software and Sovereignty.* Cambridge, Mass.: MIT Press.

Brennan, Teresa. 2004. *The Transmission of Affect.* Ithaca, N.Y.: Cornell University Press.

Briant, Emma. 2015. *Propaganda and Counter-terrorism.* Manchester: Manchester University Press.

Broad, William, and David Sanger. 2017. "U.S. Strategy to Hobble North Korea Was Hidden in Plain Sight." *New York Times,* March 4. https://www.nytimes.com/2017/03/04/world/asia/left-of-launch-missile-defense.html.

Brooking, Emerson, and P. W. Singer. 2016. "War Goes Viral: How Social Media Is Being Weaponized across the World." *Atlantic,* November. http://www.theatlantic.com/magazine/archive/2016/11/war-goes-viral/501125/.

Brooks, Rosa. 2016. *How Everything Became War and the Military Became Everything.* New York: First Simon and Schuster.

Brown, Alleen. 2017a. "Leaked Documents Reveal Counterterrorism Tactics Used at Standing Rock to 'Defeat Pipeline Insurgencies.'" *The Intercept,* May 27. https://theintercept.com/2017/05/27/leaked-documents-reveal-security-firms-counterterrorism-tactics-at-standing-rock-to-defeat-pipeline-insurgencies/.

Brown, Alleen, Will Parrish, and Alice Speri. 2017. "TigerSwan Faces Lawsuit over Unlicensed Security Operations in North Dakota." *The Intercept,* June 28. https://theintercept.com/2017/06/28/tigerswan-faces-lawsuit-over-unlicensed-security-operations-in-north-dakota/.

Brown, Barrett. 2012. "The Purpose of Project PM." *Barrett Brown* (blog), May 29. http://barrettbrown.blogspot.ca/2012/05/purpose-of-project-pm.html.

Browne, Simone. 2015. *Dark Matters: On the Surveillance of Blackness.* Durham, N.C.: Duke University Press.

Brunner, Elgin Medea. 2013. *Foreign Security Policy, Gender, and US Military Identity.* New York: Palgrave Macmillan.

Brunton, Finn, and Helen Nissenbaum. 2015. *Obfuscation: A User's Guide for Privacy and Protest.* Kindle ed. Cambridge, Mass.: MIT Press.

Bucher, Taina. 2012a. "The Friendship Assemblage: Investigating Programmed Sociality on Facebook." *Television and New Media* 14, no. 6: 479–93.

Bucher, Taina. 2012b. "A Technicity of Attention: How Software 'Makes Sense.'" *Culture Machine* 13: 1–23.

Budiansky, Stephen. 2016. *Code Warriors: NSA's Codebreakers and the Secret Intelligence War against the Soviet Union.* New York: Knopf.

Budraitskis, Ilya. 2014. "Intellectuals and the 'the New Cold War': From the Tragedy to the Farce of Choice." *LeftEast,* November 17. http://www.criticatac.ro/lefteast/intellectuals-and-the-new-cold-war/.

Budraitskis, Ilya, and Charles Davis. 2015. "'We Should Recognize That There Are Other Imperialisms': A Marxist Dissident Explains What the Left Gets Wrong about Russia." *Salon,* April 6. http://www.salon.com/2015/04/06/we_should_recognize_that_there_are_other_imperialisms_a_marxist_dissident_explains_what_the_left_gets_wrong_about_russia/.

Buhr, Sarah. 2017. "Tech Employees Protest in Front of Palantir HQ over Fears It Will Build Trump's Muslim Registry." *TechCrunch,* January 18. https://techcrunch.com/2017/01/18/tech-employees-protest-in-front-of-palantir-hq-over-fears-it-will-build-trumps-muslim-registry/.

Buncombe, Andrew. 2015. "#BlackLivesMatter Activists Were 'Monitored by Cyber Security Firm' during Baltimore Freddie Gray Protests." *Independent,* August 3. http://www.independent.co.uk/news/world/americas/blacklivesmatter-activists-were-monitored-by-cyber-security-firm-during-baltimore-freddie-gray-10435966.html.

Burman, Annie. 2013. "Gendering Decryption—Decrypting Gender: The Gender Discourse of Labour at Bletchley Park 1939–1945." MA thesis, Uppsala University.

Cadwalladr, Carole. 2017a. "British Courts May Unlock Secrets of How Trump Campaign Profiled US Voters." *Guardian,* October 1. https://www.theguardian.com/technology/2017/oct/01/cambridge-analytica-big-data-facebook-trump-voters.

Cadwalladr, Carole. 2017b. "Trump, Assange, Bannon, Farage . . . Bound Together in an Unholy Alliance." *Guardian,* October 28. https://www.theguardian.com/commentisfree/2017/oct/28/trump-assange-bannon-farage-bound-together-in-unholy-alliance.

Cadwalladr, Carole. 2018. "'I Made Steve Bannon's Psychological Warfare Tool': Meet the Data War Whistleblower." *Guardian,* March 18. https://www.theguardian.com/news/2018/mar/17/data-war-whistleblower-christopher-wylie-faceook-nix-bannon-trump.

Canabarro, Diego Rafael, and Thiago Borne. 2013. "Reflections on 'The Fog of (Cyber)War.'" Policy Working Paper 13-001, National Center for Digital Government.

Carr, Jeffrey. 2011. *Inside Cyber Warfare: Mapping the Cyber Underworld.* New York: O'Reilly.

Castells, Manuel. 2012. *Networks of Outrage and Hope: Social Movements in the Internet Age.* Cambridge: Polity.

Cederström, Carl, and Peter Fleming. 2012. *Dead Men Working*. Winchester, U.K.: Zero Books.

Chamayou, Grégoire. 2015. *A Theory of the Drone*. New York: New Press.

Chen, Adrian. 2015. "The Agency." *New York Times*, June 2. https://www.nytimes.com/2015/06/07/magazine/the-agency.html.

Chen, Adrian. 2016. "The Propaganda about Russian Propaganda." *New Yorker*, December 1. http://www.newyorker.com/news/news-desk/the-propaganda-about-russian-propaganda/.

Cheney-Lippold, John. 2017. *We Are Data: Algorithms and the Making of Our Digital Selves*. New York: New York University Press.

Chickering, Roger, Stig Förster, and Bernd Greiner, eds. 2005. *A World at Total War: Global Conflict and the Politics of Destruction, 1937–1945*. Cambridge: Cambridge University Press.

Chun, Wendy. 2004. "On Software, or the Persistence of Visual Knowledge." http://www.brown.edu/Departments/MCM/people/chun/papers/software.pdf.

Chun, Wendy. 2008. "On Sourcery, or Code as Fetish." *Configurations* 16, no. 3: 299–324.

Chun, Wendy. 2016. *Updating to Remain the Same: Habitual New Media*. Cambridge, Mass.: MIT Press.

Cimbala, Stephen J. 2017. "Nuclear Crisis Management and Deterrence: America, Russia, and the Shadow of Cyber War." *Journal of Slavic Military Studies* 30, no. 4: 487–505.

Citizen Lab. 2017. "Net Alert: Secure Your Chat." *Net Alert*, November 9. https://netalert.me/encrypted-messaging.html.

Ciuta, Felix, and Ian Klinke. 2010. "Lost in Conceptualization: Reading the 'New Cold War' with Critical Geopolitics." *Political Geography* 29: 323–32.

Clarke, Richard. 2010. *Cyber War: The Next Threat to National Security and What to Do about It*. New York: Ecco.

Cleaver, Harry. 1995. "The Zapatistas and the Electronic Fabric of Struggle." https://la.utexas.edu/users/hcleaver/zaps.html.

Cockburn, Andrew. 2015. *Kill Chain: Drones and the Rise of High-Tech Assassins*. London: Verso.

Coker, Margaret, and Paul Sonne. 2015. "Ukraine: Cyberwar's Hottest Front." *Wall Street Journal*, November 9. https://www.wsj.com/articles/ukraine-cyberwars-hottest-front-1447121671.

Coleman, Gabriella. 2015. *Hacker, Hoaxer, Whistleblower, Spy: The Many Faces of Anonymous*. With a new epilogue. London: Verso.

Coleman, Gabriella. 2017a. "From Internet Farming to Weapons of the Geek." *Current Anthropology* 58, Suppl. 15: 91–101. https://www.journals.uchicago.edu/doi/abs/10.1086/688697.

Coleman, Gabriella. 2017b. "The Public Interest Hack." *Limn* 8. http://limn.it/the-public-interest-hack/.

Collins, Brandi, Tessa D'arcangelew, Steven Renderos, and Fatima Kahn. 2015. "Stingrays, Drones and Fusion Centers: Coming to a Police Department Near You?" Netroots Nation Conference, Phoenix, Ariz., July 18. https://www.netrootsnation.org/nn_events/nn-15/stingrays-drones-and-fusion-centers-coming-to-a-police-department-near-you/.

Comninos, Alex. 2011. "Twitter Revolutions and Cyber Crackdowns: User-Generated Content and Social Networking in the Arab Spring and Beyond." Association for Progressive Communications. https://www.apc.org/sites/default/files/AlexComninos_MobileInternet.pdf.

Conger, Kate. 2018a. "Amazon Workers Demand Jeff Bezos Cancel Face Recognition Contracts with Law Enforcement." Gizmodo, June 21. https://gizmodo.com/amazon-workers-demand-jeff-bezos-cancel-face-recognitio-1827037509.

Conger, Kate. 2018b. "Google Employees Resign in Protest against Pentagon Contract." Gizmodo, May 14. https://gizmodo.com/google-employees-resign-in-protest-against-pentagon-con-1825729300.

Cordesman, Anthony. 2014. "Russia and the Color Revolution." Centre for Strategic and International Studies, May 28. http://csis.org/publication/russia-and-color-revolution/.

Corera, Gordon. 2015. "How France's TV5 Was Almost Destroyed by 'Russian hackers.'" BBC News, October 10. http://www.bbc.com/news/technology-37590375.

Crisp, James. 2017. "EU Governments to Warn Cyber Attacks Can Be an Act of War." Telegraph, October 29. http://www.telegraph.co.uk/news/2017/10/29/eu-governments-warn-cyber-attacks-can-act-war/.

Cronin, Audrey. 2006. "Cyber-Mobilization: The New Levée en Masse." Parameters 36, no. 2: 77–87.

Cuthbertson, Anthony. 2016. "Isis Launches Cyberwar Magazine for Wannabe Jihadist Hackers." International Business Observer, January 6. http://www.ibtimes.co.uk/isis-launches-cyberwar-magazine-jihadists-making-1536334/.

Damon, Andre. 2017. "Google Blocked Every One of the WSWS's 45 Top Search Terms." World Socialist Web Site. August 4. https://www.wsws.org/en/articles/2017/08/04/goog-a04.html.

Damon, Andre, and David North. 2017. "Google's New Search Protocol Is Restricting Access to 13 Leading Socialist, Progressive and Anti-war Web Sites." World Socialist Web Site. August 2. https://www.wsws.org/en/articles/2017/08/02/pers-a02.html.

Darczewska, Jolanta. 2014a. "The Anatomy of Russian Information Warfare the Crimean Operation, a Case Study." Point of View 42. http://www.osw.waw.pl/sites/default/files/the_anatomy_of_russian_information_warfare.pdf.

Darczewska, Jolanta. 2014b. The Information War on Ukraine: New Challenges. Cicero Foundation Great Debate Paper 14/08. http://www.cicerofoundation.org/lectures/Jolanta_Darczewska_Info_War_Ukraine.pdf.

Davis, Jim, Thomas Hirschl, and Michael Stack, eds. 1997. *Cutting Edge: Technology, Information, Capitalism and Social Revolution.* London: Verso.

Dean, Jodi. 2005. "Communicative Capitalism: Circulation and the Foreclosure of Politics." *Cultural Politics* 1, no. 1: 51–74.

Dean, Jodi. 2008. "Enjoying Neoliberalism." *Cultural Politics* 4, no. 1: 47–72.

Dean, Jodi. 2009. *Democracy and Other Neoliberal Fantasies: Communicative Capitalism and Left Politics.* Durham, N.C.: Duke University Press.

Dean, Jodi. 2010. *Blog Theory: Feedback and Capture in the Circuits of Drive.* Cambridge: Polity.

Dean, Jodi. 2012. *The Communist Horizon.* London: Verso.

Dean, Jodi. 2014. "The Real Internet." In *Žižek and Media Studies: A Reader,* edited by Matthew Flisfeder and Louis-Paul Willis, 211–28. Houndsville, U.K.: Palgrave Macmillan.

De Gaudemar, Jean-Paul. 1979. *La Mobilisation Générale.* Paris: Gallica.

Deibert, Ronald. 2013. *Black Code: Inside the Battle for Cyberspace.* Toronto, Ont.: Signal, McClelland, and Stewart.

Deibert, Ronald. 2015. "The Geopolitics of Cyberspace after Snowden." *Current History* 114: 9–15. http://www.currenthistory.com/Deibert_Current History.pdf.

Deibert, Ronald, Rafal Rohozinski, and Masashi Crete-Nishihata. 2012. "Cyclones in Cyberspace: Information Shaping and Denial in the 2008 Russia–Georgia War." *Security Dialogue* 43, no. 1: 3–24.

DeLanda, Manuel. 1991. *War in the Age of Intelligent Machines.* New York: Zone Books.

Deleuze, Giles. 1992. "Postscript on the Societies of Control." *October* 59: 3–7.

Deleuze, Gilles, and Félix Guattari. 1986. *Nomadology: The War Machine.* New York: Semiotexte.

Dencik, Lina, Arne Hintz, and Jonathan Cable. 2016. "Towards Data Justice? The Ambiguity of Anti-surveillance Resistance in Political Activism." *Big Data and Society* 3, no. 2: 1–12.

Deseriis, Marco. 2017. "Hacktivism: On the Use of Botnets in Cyberattacks." *Theory, Culture, and Society* 34, no. 4: 131–52.

Deterritorial Support Group. 2012. "All the Memes of Production." New Left Project. http://www.newleftproject.org/index.php/site/article_comments /all_the_memes_of_production.

DeYoung, Karen, Ellen Nakashima, and Emily Rauhala. 2017. "Trump Signed Presidential Directive Ordering Actions to Pressure North Korea." *Washington Post,* September 30. https://www.washingtonpost.com/world/national -security/trump-signed-presidential-directive-ordering-actions-to-pressure -north-korea/2017/09/30/97c6722a-a620-11e7-b14f-f41773cd5a14_story.html.

Dolar, Mladen. 1993. "Beyond Interpellation." *Qui Parle* 6, no. 2: 75–96. http:// www.jstor.org/stable/20685977/.

Dolcourt, Jessica. 2017. "By 2021, Most Internet Devices Won't Be for Humans." *C/Net*, June 8. https://www.cnet.com/news/by-2021-most-internet-devices-wont-be-for-humans/.

Doshi, Vidhi. 2017. "India's Millions of New Internet Users Are Falling for Fake News—Sometimes with Deadly Consequences." *Washington Post*, October 1. https://tinyurl.com/y8gffufk.

Draper, Hal, and Ernest Haberkern. 2005. *Karl Marx's Theory of Revolution: Vol. V. War and Revolution*. New York: Monthly Review Press.

Dreiziger, N. F., ed. 2006. *Mobilization for Total War: The Canadian, American and British Experience 1914–18 and 1939–45*. Waterloo: Wilfred Laurier.

Dunlap, Charles, Jr. 2014. "The Hyper-personalization of War: Cyber, Big Data, and the Changing Face of Conflict." *Georgetown Journal of International Affairs*. https://scholarship.law.duke.edu/cgi/viewcontent.cgi?article=6068&context=faculty_scholarship/.

Dyer-Witheford, Nick. 1999. *Cyber-Marx: Circuits and Cycles of Struggle in High-Technology Capitalism*. Champagne: University of Illinois Press.

Dyer-Witheford, Nick. 2015. *Cyber-Proletariat: Global Labour in the Digital Vortex*. London: Pluto Press.

Dzarasov, Ruslan. 2014. *The Conundrum of Russian Capitalism: The Post-Soviet Economy in the World System*. London: Pluto Press.

Economist. 2015a. "The IS Media Machine." October 8. https://www.economist.com/blogs/graphicdetail/2015/10/tracking-islamic-states-media-output/.

Economist. 2015b. "The Propaganda War." August 15. https://www.economist.com/news/middle-east-and-africa/21660989-terrorists-vicious-message-surprisingly-hard-rebut-propaganda-war/.

Economist. 2016a. "China Invents the Digital Totalitarian State." December 17. https://www.economist.com/news/briefing/21711902-worrying-implications-its-social-credit-project-china-invents-digital-totalitarian/.

Economist. 2016b. "The Languages of Jihad." February 16. https://www.economist.com/news/international/21571867-islamic-extremists-are-increasingly-multilingual-bunch-especially-online-languages/.

Economist. 2017a. "Military Robots Are Getting Smaller and More Capable." December 14. https://www.economist.com/news/science-and-technology/21732507-soon-they-will-travel-swarms-military-robots-are-getting-smaller-and-more.

Economist. 2017b. "Once Considered a Boon to Democracy, Social Media Have Started to Look Like Its Nemesis." October 4. https://www.economist.com/news/briefing/21730870-economy-based-attention-easily-gamed-once-considered-boon-democracy-social-media.

Economist. 2017c. "Out of VKontakte: Ukraine Bans Its Top Social Networks because They Are Russian." May 17. https://www.economist.com/news/europe/21722360-blocking-websites-may-be-pointless-it-could-help-president-poroshenkos-popularity-ukraine.

Economist. 2018. "The New Battlegrounds." Special Supplement 42, January 27.

Edwards, Paul. 1996. *The Closed World: Computers and the Politics of Discourse in Cold War America.* Cambridge, Mass.: MIT Press.

Egan, Daniel. 2013. *The Dialectic of Position and Maneuver: Understanding Gramsci's Military Metaphor.* Leiden, Netherlands: Brill.

Elmer, Greg. 2004. *Profiling Machines: Mapping the Personal Information Economy.* Cambridge, Mass.: MIT Press.

Emmons, Alex. 2018. "With Support from Nancy Pelosi, House Gives Trump Administration Broad Latitude to Spy on Americans." *Intercept,* January 11. https://theintercept.com/2018/01/11/nsa-pelosi-democrats-spy-american-section-702/.

Emridge, Julain. 2017. "The Forgotten Interventions." *Jacobin,* January 12. https://www.jacobinmag.com/2017/01/russia-hacks-election-meddling-iran-mossadegh-chile-allende-guatemala-arbenz-coup/.

Enloe, Cynthia. 2016. *Globalization and Militarism: Feminists Make the Link.* Lanham, Md.: Rowman and Littlefield.

Entous, Adam, Ellen Nakashima, and Greg Miller. 2016. "Secret CIA Assessment Says Russia Was Trying to Help Trump Win White House." *Washington Post,* December 9. https://www.washingtonpost.com/world/national-security/obama-orders-review-of-russian-hacking-during-presidential-campaign/2016/12/09/31d6b300-be2a-11e6-94ac-3d324840106c_story.html.

Equality Labs. 2017. "Digital Self Defense in the Time of Trump." Equality Labs. https://www.equalitylabs.org/.

Farmer, Ben. 2014. "Ukraine Cyber War Escalates alongside Violence." *Telegraph,* May 28. http://www.telegraph.co.uk/news/worldnews/europe/ukraine/10860920/Ukraine-cyber-war-escalates-alongside-violence.html.

Fattor, Eric M. 2014. *American Empire and the Arsenal of Entertainment: Soft Power and Cultural Weaponization.* New York: Palgrave Macmillan.

Fielding, Nick, and Ian Cobain. 2011. "Revealed: US Spy Operation That Manipulates Social Media." *Guardian,* March 17. https://www.theguardian.com/technology/2011/mar/17/us-spy-operation-social-networks/.

Filiol, Eric. 2011. "Operational Aspects of Cyberattack: Intelligence, Planning and Conduct." In *Cyberwar and Information Warfare,* edited by Daniel Ventre, 255–84. Hoboken, N.J.: John Wiley.

Finn, Ed. 2017. *What Algorithms Want: Imagination in the Age of Computing.* Cambridge, Mass.: MIT Press.

Flisfeder, Matthew. 2014. "Enjoying Social Media." In *Žižek and Media Studies: A Reader,* edited by Matthew Flisfeder and Louis-Pau Willis, 229–40. Houndsville, U.K.: Palgrave Macmillan.

Flisfeder, Matthew. 2018. "The Ideological Algorithmic Apparatus: Subjection before Enslavement." *Theory and Event* 21, no. 2: 457–84.

Foer, Franklin. 2017. *World without Mind: The Existential Threat of Big Tech.* New York: Penguin Press.

Franceschi-Bicchierai, L. 2016. "Hacker 'Phineas Fisher' Speaks on Camera for the First Time—through a Puppet." *Motherboard*, July 20. https://mother board.vice.com/en_us/article/78kwke/hacker-phineas-fisher-hacking -team-puppet.

Fraser, Nancy. 2013. *Fortunes of Feminism: From State-Managed Capitalism to Neoliberal Crisis*. London: Verso.

Frenkel, Shara. 2018. "Microsoft Employees Protest Work with ICE, as Tech Industry Mobilizes over Immigration." *New York Times*, June 19. https:// www.nytimes.com/2018/06/19/technology/tech-companies-immigration -border.html.

Freud, Sigmund. 1900. "The Interpretation of Dreams (Second Part) (1900)." *Standard Edition* 5: 339–627.

Fuchs, Christian. 2014. *Social Media: A Critical Introduction*. Los Angeles, Calif.: Sage.

Furst, Alan. 1988. *Night Soldiers*. New York: Random House.

Furst, Alan. 1991. *Dark Star*. New York: Houghton Mifflin.

Galeotti, Mark. 2014. "The 'Gerasimov Doctrine' and Russian Non-linear War." *Moscow's Shadows*, July 6. https://inmoscowsshadows.wordpress.com /2014/07/06/the-gerasimov-doctrine-and-russian-non-linear-war/.

Galeotti, Mark. 2018. "The Mythical 'Gerasimov Doctrine.'" *Critical Studies on Security*. https://doi.org/10.1080/21624887.2018.1441623.

Gallagher, Ryan. 2018. "Google Plans to Launch Censored Search Engine in China, Leaked Documents Reveal." *The Intercept* (blog), August 1. https:// theintercept.com/2018/08/01/google-china-search-engine-censorship/.

Gallagher, Ryan, and Glen Greenwald. 2014. "How the NSA Plans to Infect 'Millions' of Computers with Malware." *The Intercept* (blog), March 14. https://theintercept.com/2014/03/12/nsa-plans-infect-millions-computers -malware/.

Gallagher, Sean. 2018a. "How They Did It (and Will Likely Try Again): GRU Hackers vs. US Elections." *Ars Technica*, July 27. https://arstechnica.com /information-technology/2018/07/from-bitly-to-x-agent-how-gru-hackers -targeted-the-2016-presidential-election/.

Gallagher, Sean. 2018b. "New Data Shows China Has 'Taken the Gloves Off' in Hacking Attacks on US." *Ars Technica*, November 1. https://arstechnica .com/information-technology/2018/11/new-data-shows-china-has-taken -the-gloves-off-in-hacking-attacks-on-us/.

Gallagher, Sean. 2018c. "Twitter 'Bot' Purge Causes Outcry from Trollerati as Follower Counts Fall." *ArsTechnica*, February 22. https://arstechnica .com/tech-policy/2018/02/twitter-suspends-thousands-of-accounts-for -bot-behavior-some-cry-censorship/.

Gartzke, Erik, and Jon R. Lindsay. 2017. "Thermonuclear Cyberwar." *Journal of Cybersecurity* 3, no. 1: 37–48.

Gellman, Barton, and Ellen Nakashima. 2013. "US Spy Agencies Mounted 23

Offensive Cyber-operations in 2011, Documents Show." *Washington Post,* August 31. https://www.washingtonpost.com/world/national-security /us-spy-agencies-mounted-231-offensive-cyber-operations-in-2011-docu ments-show/2013/08/30/do90a6ae-119e-11e3-b4cb-fd7ceo41d814_story.html.

Gerasimov, Valery. 2013. "The Value of Science Is in the Foresight: New Challenges Demand Rethinking the Forms and Methods of Carrying Out Combat Operations." *Military-Industrial Kurier,* February 27. https://us acac.army.mil/CAC2/MilitaryReview/Archives/English/MilitaryReview _20160228_art008.pdf.

Gerbaudo, Paolo. 2012. *Tweets and the Streets: Social Media and Contemporary Activism.* London: Pluto.

Gertz, Bill. 2017. *iWar: War and Peace in the Information Age.* New York: Simon and Schuster.

Gibson, William. 1984. *Neuromancer.* New York: ACE.

Gibson, William. 2015. *The Peripheral.* New York: Berkley.

Gill, Tim. 2017. "It Didn't End with the Cold War." *Jacobin,* January 25. https:// www.jacobinmag.com/2017/01/us-intervention-russia-elections-imperial ism-latin-america/.

Gjelten, Tom. 2013. "Pentagon Goes on the Offensive against Cyberattacks." NPR. February 11. https://www.npr.org/2013/02/11/171677247/pentagon -goes-on-the-offensive-against-cyber-attacks/.

Glaser, April. 2018. "The Incomplete Vision of John Perry Barlow." *Slate,* February 8. https://slate.com/technology/2018/02/john-perry-barlow-gave -internet-activists-only-half-the-mission-they-need.html.

Gongora, Thierry, and Harald von Riekhoff, eds. 2000. *Toward a Revolution in Military Affairs? Defense and Security at the Dawn of the Twenty-First Century.* Westport, Conn.: Greenwood Press.

Goodin, Dan. 2015. "Meet 'Great Cannon,' the Man-in-the-Middle Weapon China Used on GitHub." *Ars Technica,* October 4. https://arstechnica .com/information-technology/2015/04/meet-great-cannon-the-man-in -the-middle-weapon-china-used-on-github/.

Goodin, Dan. 2017. "Russia Reportedly Stole NSA Secrets with Help of Kaspersky—What We Know Now." *Ars Technica,* October 5. https://ars technica.com/information-technology/2017/10/the-cases-for-and-against -claims-kaspersky-helped-steal-secret-nsa-secrets/.

Goodin, Dan. 2018. "12 Russian Intel Officers Indicted for Hacking the DNC and Clinton Campaign." *Ars Technica,* July 13. https://arstechnica.com /tech-policy/2018/07/12-russian-intel-officers-indicted-for-hacking-the -dnc-and-clinton-campaign/.

Gorwa, Robert. 2017. "Twitter Has a Serious Bot Problem, and Wikipedia Might Have the Solution." *Quartz,* October 21. https://qz.com/1108092 /twitter-has-a-serious-bot-problem-and-wikipedia-might-have-the-solu tion/.

Graeber, David. 2016. Foreword to *Revolution in Rojava: Democratic Autonomy and Women's Liberation in Syrian Kurdistan,* edited by Michael Knapp et al., xii–xxii. London: Pluto Press.

Graham, Stephen. 2016. *Vertical: The City from Satellites to Bunkers.* New York: Verso.

Gray, Bill. 2017. "Review: The Operational Art of War IV." *Wargamer,* November 16. http://www.wargamer.com/reviews/review-the-operational-art-of-war-iv/.

Gray, Chris Hables, and Ángel J. Gordo. 2014. "Social Media in Conflict Comparing Military and Social-Movement Technocultures." *Cultural Politics* 10, no. 3: 251–61.

Gray, Colin S. 2004. *Strategy for Chaos: Revolutions in Military Affairs and the Evidence of History.* London: Frank Cass.

Green, Kieran Richard. 2016. "People's War in Cyberspace: Using China's Civilian Economy in the Information Domain." *Military Cyber Affairs* 2, no. 1. http://scholarcommons.usf.edu/mca/vol2/iss1/5/.

Greenberg, Andrew. 2012. *This Machine Kills Secrets: How Wikileakers, Cypherpunks and Hacktivists Aim to Free the World's Information.* New York: Dutton.

Greenberg, Andrew. 2015. "A Second Snowden Has Leaked Motherlode of Drone Documents." *Wired,* October 15. https://www.wired.com/2015/10/a-second-snowden-leaks-a-mother-lode-of-drone-docs/.

Greenberg, Andrew. 2017a. "How an Anarchist Bitcoin Coder Found Himself Fighting ISIS in Syria." *Wired,* March 29. https://www.wired.com/2017/03/anarchist-bitcoin-coder-found-fighting-isis-syria/.

Greenberg, Andrew. 2017b. "How an Entire Nation Became Russia's Test Lab for Cyberwar." *Wired,* June 20. https://www.wired.com/story/russian-hackers-attack-ukraine/.

Greenwald, Glenn. 2013. *No Place to Hide: Edward Snowden, the NSA, and the U.S. Surveillance State.* New York: Metropolitan Books.

Greenwald, Glenn. 2016. "New Study Shows Mass Surveillance Breeds Meekness, Fear and Self-Censorship." *Intercept,* April 28. https://theintercept.com/2016/04/28/new-study-shows-mass-surveillance-breeds-meekness-fear-and-self-censorship/.

Greenwald, Glenn. 2017a. "CNN Journalists Resign: Latest Example of Media Recklessness on the Russia Threat." *Intercept,* June 7. https://theintercept.com/2017/06/27/cnn-journalists-resign-latest-example-of-media-recklessness-on-the-russia-threat/.

Greenwald, Glenn. 2017b. "Facebook Says It Is Deleting Accounts at the Direction of the U.S. and Israeli Governments." *Intercept,* December 30. https://theintercept.com/2017/12/30/facebook-says-it-is-deleting-accounts-at-the-direction-of-the-u-s-and-israeli-governments/.

Greenwald, Glenn. 2017c. "Yet Another Major Russia Story Falls Apart. Is

Skepticism Permissible Yet?" *Intercept,* September 28. https://theintercept
.com/2017/09/28/yet-another-major-russia-story-falls-apart-is-skepticism
-permissible-yet/.

Greenwald, Glenn. 2018. "Dutch Official Admits Lying about Meeting with
Putin: Is Fake News Used by Russia or about Russia?" *Intercept,* February
12. https://theintercept.com/2018/02/12/dutch-official-admits-lying-about
-meeting-with-putin-is-fake-news-used-by-russia-or-about-russia/.

Gregg, Aaron. 2017. "Amazon Launches New Cloud Storage Service for U.S. Spy
Agencies." *Washington Post,* November 20. https://www.washingtonpost
.com/news/business/wp/2017/11/20/amazon-launches-new-cloud-storage
-service-for-u-s-spy-agencies/.

Gruzd, Anatoliy, and Ksenia Tsyganova. 2014. "Politically Polarized Online
Groups and Their Social Structures Formed around the 2013–2014 Crisis
in Ukraine." Paper presented at the Internet, Policy and Politics (IPP)
Conference: Crowdsourcing for Politics and Policy, University of Oxford,
September 25–26. http://ipp.oii.ox.ac.uk/sites/ipp/files/documents/IPP
2014_Gruzd.pdf.

Hafner, Katie. 1998. *Where Wizards Stay Up Late: The Origins of the Internet.* New
York: Simon and Schuster.

Hahn, Gordon M. 2017. *Ukraine over the Edge: Russia, the West and the "New Cold
War."* New York: McFarlane.

Hambling, David. 2009. "China Looks to Undermine U.S. Power, with 'Assassin's
Mace.'" *Wired,* July 2. https://www.wired.com/2009/07/china-looks-to
-undermine-us-power-with-assassins-mace/.

Hands, Joss. 2011. *@ Is for Activism: Dissent, Resistance and Rebellion in a Digital
Culture.* London: Pluto.

Hansen, Lene, and Helen Nissenbaum. 2009. "Digital Disaster, Cybersecurity
and the Copenhagen School." *International Studies Quarterly* 53: 1155–75.

Harding, Luc. 2016. "Ukraine's Leader Set Up Secret Offshore Firm as Battle
Raged with Russia." *Guardian,* April 4. https://www.theguardian.com
/news/2016/apr/04/panama-papers-ukraine-petro-poroshenko-secret
-offshore-firm-russia/.

Harding, Luke, and Andrew Roth. 2018. "Arkady Babchenko Reveals He Faked
His Death to Thwart Moscow Plot." *Guardian,* May 30. https://www
.theguardian.com/world/2018/may/30/arkady-babchenko-reveals-he-faked
-his-death-to-thwart-moscow-plot.

Hardt, Michael, and Antonio Negri. 2000. *Empire.* Cambridge, Mass.: Harvard
University Press.

Hardt, Michael, and Antonio Negri. 2009. *Commonwealth.* Boston: Belknap Press.

Hardt, Michael, and Antonio Negri. 2017. *Assembly.* Oxford: Oxford University
Press.

Harris, Lalage. 2017. "Letter from Velles: The Real Story of Macedonia's Fake

News Factory." https://www.calvertjournal.com/features/show/8031/letter-from-veles/.

Harris, Shane. 2014. @War: The Rise of the Military–Internet Complex. New York: Houghton Mifflin.

Hart-Landsberg, Martin. 2013. Capitalist Globalization: Consequences, Resistance, and Alternatives. New York: Monthly Review.

Healey, Jason, ed. 2013. A Fierce Domain: Conflict in Cyberspace, 1986–2012. New York: CCSA/Atlantic Council.

Herman, Edward S., and Noam Chomsky. 2002. Manufacturing Consent: The Political Economy of the Mass Media. New York: Pantheon Books.

Herrera, Linda. 2014. Revolution in the Age of Social Media: The Egyptian Popular Insurrection and the Internet. London: Verso.

Hersh, Seymour M. 2007. "The Redirection." New Yorker, March 5. http://www.newyorker.com/magazine/2007/03/05/the-redirection/.

Hersh, Seymour M. 2010. "The Online Threat: Should We Be Worried about a Cyber War?" New Yorker, October 25. http://www.newyorker.com/magazine/2010/11/01/the-online-threat.

Hicks, Marie. 2018. Programmed Inequality: How Britain Discarded Women Technologists and Lost Its Edge in Computing. Cambridge, Mass.: MIT Press.

Himanen, Pekka. 2002. The Hacker Ethic and the Spirit of the Information Age. New York: Random House.

Hoffman, Bruce. 2008. "The Myth of Grass-Roots Terrorism: Why Osama bin Laden Still Matters." Foreign Affairs, May/June. https://www.foreignaffairs.com/reviews/review-essay/2008-05-03/myth-grass-roots-terrorism.

Hoffman, Frank. 2013. "You May Not Be Interested in Cyber War, but It Is Interested in You. War on the Rocks (blog), August 7. https://warontherocks.com/2013/08/you-may-not-be-interested-in-cyber-war-but-its-interested-in-you/.

Hsu, Jeremy. 2014. "U.S. Suspicions of China's Huawei Based Partly on NSA's Own Spy Tricks." IEEE Spectrum, March 26. https://spectrum.ieee.org/tech-talk/computing/hardware/us-suspicions-of-chinas-huawei-based-partly-on-nsas-own-spy-tricks/.

Hung, Ho-Fung. 2009. "America's Head Servant? The PRC's Dilemma in the Global Crisis." New Left Review 60: 3–25.

Illing, Sean. 2017. "Cambridge Analytica, the Shady Data Firm That Might Be a Key Trump-Russia Link, Explained." Vox, December 18. https://www.vox.com/policy-and-politics/2017/10/16/15657512/mueller-fbi-cambridge-analytica-trump-russia.

Invisible Committee. 2015. To Our Friends. New York: Semiotext(e).

Ioffe, Julia. 2017. "The Secret Correspondence between Donald Trump Jr. and WikiLeaks." Atlantic, November 13. https://www.theatlantic.com/politics/archive/2017/11/the-secret-correspondence-between-donald-trump-jr-and-wikileaks/545738/.

Ismail, Nick. 2017. "University Cyber Security Breaches Double in 2 Years and Leak Military Secrets." *Information Age,* September 5. http://www.information-age.com/university-cyber-security-breaches-leak-military-secrets-123468364/.

Jameson, Frederic. 2002. "The Dialectics of Disaster." *South Atlantic Quarterly* 101, no. 2: 297–304.

Jamieson, Kathleen Hall. 2018. *Cyberwar: How Russian Hackers and Trolls Helped Elect a President—What We Don't, Can't, and Do Know.* Oxford: Oxford University Press.

Johnson, Henry. 2015. "Kurdish Fighters: We'd Really Appreciate If You Came to Syria to Help Fight ISIS." *Foreign Policy,* September 17. http://foreignpolicy.com/2015/09/17/kurdish-fighters-wed-really-appreciate-if-you-came-to-syria-to-help-fight-isis/.

Johnston, John. 2008. *The Allure of Machinic Life: Cybernetics, Artificial Life and the New AI.* Cambridge, Mass.: MIT Press.

Jones, Andrew Jerell. 2014. "Russian Interest in Ukraine Now Includes Cyber Warfare." *Intercept,* October 17. https://firstlook.org/theintercept/2014/10/17/russian-hackers-find-flaw-microsoft-windows-spy/.

Jones, Marc. 2017. "Hacking, Bots and Information Wars in the Qatar Spat." *Washington Post,* June 7. https://www.washingtonpost.com/news/monkey-cage/wp/2017/06/07/hacking-bots-and-information-wars-in-the-qatar-spat/.

Jones, Sam. 2016. "Cyber-espionage: A New Cold War." *Financial Times,* August 19. https://www.ft.com/content/d63c5b3a-65ff-11e6-a08a-c7ac04efooaa.

Joque, Justin. 2018. *Deconstruction Machines: Writing in the Age of Cyberwar.* Kindle ed. Minneapolis: University of Minnesota Press.

Jordan, Tim. 2008. *Hacking Digital Media and Technological Activism.* Cambridge: Polity.

Jordan, Tim. 2015. *Information Politics: Liberation and Exploitation in the Digital Society.* London: Pluto Press.

Jordan, Tim, and Paul A. Taylor. 2004. *Hacktivism and Cyberwars: Rebels with a Cause?* New York: Routledge.

Joseph, George. 2015. "Exclusive: Feds Regularly Monitored Black Lives Matter since Ferguson." *Intercept,* June 24. https://theintercept.com/2015/07/24/documents-show-department-homeland-security-monitoring-black-lives-matter-since-ferguson/.

Joseph, George. 2017. "NYPD officers accessed Black Lives Matter activists' texts, documents show." *Guardian,* April 4. https://www.theguardian.com/us-news/2017/apr/04/nypd-police-black-lives-matter-surveillance-undercover.

Just, Natascha, and Michael Latzer. 2017. "Governance by Algorithms: Reality Construction by Algorithmic Selection on the Internet." *Media, Culture, and Society* 39, no. 2: 238–58.

Kaiser, Robert. 2015. "The Birth of Cyberwar." *Political Geography* 46: 11–20.

Kaplan, Fred. 2016. *Dark Territory: The Secret History of Cyberwar.* New York: Simon and Schuster.

Kaplan, Fred. 2018. "Nuclear Posturing." *Slate,* January 22. https://slate.com/news-and-politics/2018/01/trumps-official-nuclear-policy-reaffirms-the-terrifying-status-quo.html.

Karatzogianni, Athina. 2015. *Firebrand Waves of Digital Activism, 1994–2014: The Rise and Spread of Hacktivism and Cyberconflict.* New York: Palgrave Macmillan.

Kastrenakes, Jacob. 2018. "Trump Signs Bill Banning Government Use of Huawei and ZTE Tech." *Verge,* August 13. https://www.theverge.com/2018/8/13/17686310/huawei-zte-us-government-contractor-ban-trump.

Keizer, Gregg. 2010. "Botnets 'the Swiss Army Knife of Attack Tools.'" *CSO,* April 7. https://www.csoonline.com/article/2124994/malware-cybercrime/botnets—the-swiss-army-knife-of-attack-tools-.html.

Kenney, Michael. 2015. "Cyber-Terrorism in a Post-Stuxnet World." *Orbis* 59, no. 1: 110–28.

Kerschischnig, Georg. 2012. *Cyberthreats and International Law.* The Hague: Eleven International.

Khatchadourian, Raffi. 2017. "Julian Assange: A Man without a Country, WikiLeaks, and the 2016 Presidential Election." *New Yorker,* August 21. https://www.newyorker.com/magazine/2017/08/21/julian-assange-a-man-without-a-country/.

King-Close, Alexandria Marie. 2016. "A Gender Analysis of Cyber War." MA thesis, Harvard Extension School. http://nrs.harvard.edu/urn-3:HUL.InstRepos:33797321.

Kipp, Jacob. 1985. "Lenin and Clausewitz: The Militarization of Marxism 1914–21." *Military Affairs* 49, no. 4: 184–91.

Kirby, Emma Jane. 2016. "The City Getting Rich from Fake News." *BBC News,* December 5. http://www.bbc.com/news/magazine-38168281/.

Kirschgaessner, Stephanie. 2017. "Cambridge Analytica Used Data from Facebook and Politico to Help Trump." *Guardian,* November 26. https://www.theguardian.com/technology/2017/oct/26/cambridge-analytica-used-data-from-facebook-and-politico-to-help-trump/.

Kittler, Friedrich. 1997. "The World of the Symbolic—a World of the Machine." In *Literature, Media, Information Systems: Essays,* edited by John Johnston, 130–46. Amsterdam: G+B Arts International.

Klimburg, Alexander. 2017. *The Darkening Web: The War for Cyberspace.* New York: Penguin Press.

Kline, Steven, Nick Dyer-Witheford, and Greig de Peuter. 2003. *Digital Play: Technology, Markets and Culture.* Montreal, Quebec: McGill-Queens.

Knapp, Michael, Anja Flac, and Asya Abdullah, eds. 2016. *Revolution in Rojava: Democratic Autonomy and Women's Liberation in Syrian Kurdistan.* London: Pluto Press.

Knapton, Sarah. 2017. "Home Office Blames North Korea for Devastating NHS 'WannaCry' Cyber Attack." *Telegraph,* October 27. http://www.telegraph .co.uk/science/2017/10/27/home-office-blames-north-korea-devastating -nhs-wannacry-cyber/.

Knibbs, Kate. 2015. "Surprise! The NSA Is Still Spying on You." *Gizmodo,* November 30. https://gizmodo.com/surprise-the-nsa-is-still-spying-on -you-1745256761.

Knox, MacGregor, and Williamson Murray, ed. 2001. *The Dynamics of Military Revolution, 1300–2050.* Cambridge: Cambridge University Press.

Kofman, Ava. 2018. "Can #MeToo Change the Toxic Culture of Sexism and Harassment at Cybersecurity Conferences?" *The Intercept* (blog), June 19. https://theintercept.com/2018/06/19/metoo-cybersecurity-infosec-sexual -harassment/.

Kopfstein, Joseph. 2016. "NSA's Hacker-in-Chief: We Don't Need Zero-Days to Get Inside Your Network." *Motherboard,* January 29. https://motherboard .vice.com/en_us/article/wnx5bm/nsas-hacker-in-chief-we-dont-need-zero -days-to-get-inside-your-network-rob-boyce.

Kopfstein, Joseph. 2017. "Want to Hide from Face Recognition? Try an Anti-Surveillance T-Shirt." *Vocativ,* January 4. http://www.vocativ.com/389682 /anti-surveillance-clothing-face-recognition/index.html.

Krautwurst, Udo. 2007. "Cyborg Anthropology and/as Endocolonisation." *Culture, Theory, and Critique* 48, no. 2: 139–60.

Kravets, David. 2016. "The Intercept Releasing Docs Leaked by NSA Whistle-blower Snowden." *Intercept,* May 16. http://arstechnica.com/tech-policy /2016/05/the-intercept-releasing-docs-leaked-by-nsa-whistleblower -snowden/.

Krebs, Brian. 2016. "Researchers Find Fresh Fodder for IoT Attack Cannons." *Krebs on Security* (blog), December 6. https://krebsonsecurity.com/2016/12 /researchers-find-fresh-fodder-for-iot-attack-cannons/.

Küçük, Bülent, and Ceren Özselçuk. 2016. "The Rojava Experience: Possibilities and Challenges of Building a Democratic Life." *South Atlantic Quarterly* 115, no. 1: 184–96.

Kundnani, Arun, and Deepa Kumar. 2015. "Race, Surveillance, and Empire." *International Socialist Review* 96. https://isreview.org/issue/96/race-sur veillance-and-empire.

Kurz, Robert. (1997) 2011. "The Destructive Origins of Capitalism." July 14. https://libcom.org/history/destructive-origins-capitalism-robert-kurz/.

Lacan, Jacques. (1967) 1997. *The Seminar, Book III: The Psychoses, 1955–56.* Translated by Russell Grigg. New York: W. W. Norton.

Lacan, Jacques. (1969–70) 2007. *The Seminar of Jacques Lacan, Book XVII: The Other Side of Psychoanalysis, 1969–70.* Translated by Russell Grigg. New York: W. W. Norton.

Lacan, Jacques. 1998. *The Seminar of Jacques Lacan, Book XI: The Four Fundamental*

Concepts of Psychoanalysis. Edited by Jacques-Alain Miller. Translated by Alan Sheridan. New York: W. W. Norton.

Lacan, Jacques. 1999. *The Seminar, Book XXI: Encore. On Feminine Sexuality the Limits of Love and Knowledge.* Translated by Bruce Fink. New York: W. W. Norton.

Lacan, Jacques. 2016. *The Seminar of Jacques Lacan, Book X: Anxiety.* London: Polity.

Lamonthe, Dan. 2017. "How the Pentagon's Cyber Offensive against ISIS Could Shape the Future for Elite U.S. Forces." *Washington Post,* December 10. https://www.washingtonpost.com/news/checkpoint/wp/2017/12/16/how-the-pentagons-cyber-offensive-against-isis-could-shape-the-future-for-elite-u-s-forces/.

Lane, David. 2014. *The Capitalist Transformation of State Socialism.* London: Routledge.

Langlois, Ganaele, Fenwick McKelvey, Greg Elmer, and Kenneth Werbin. 2009. "Mapping Commercial Web 2.0 Worlds: Towards a New Critical Ontogenesis." *Fibreculture Journal* 14. http://fourteen.fibreculturejournal.org/fcj-095-mapping-commercial-web-2-0-worlds-towards-a-new-critical-ontogenesis/.

Lapowsky, Issie. 2018. "Facebook Exposed 87 Million Users to Cambridge Analytica." *Wired,* April 4. https://www.wired.com/story/facebook-exposed-87-million-users-to-cambridge-analytica/.

Lecher, Colin. 2018. "House Votes to Renew Controversial Surveillance Program That Powers the NSA." *Verge,* January 11. https://www.theverge.com/2018/1/11/16874158/house-702-fisa-bill-vote-nsa-surveillance-spying.

Leezenberg, Michael. 2016. "The Ambiguities of Democratic Autonomy: The Kurdish Movement in Turkey and Rojava." *Southeast Europe and Black Sea Studies* 16, no. 4: 671–90.

Lenin, Vladymyr Illich. (1902) 1969. *What Is to Be Done? Burning Questions of Our Movement.* Moscow: International.

Lenin, Vladimir Illich. (1917) 1964. "War and Revolution: A Lecture Delivered May 14, 1917." In *Lenin Collected Works,* 24:398–421. Moscow: Progress. https://www.marxists.org/archive/lenin/works/1917/may/14.htm.

Lenin, Vladimir Illich. 1939. *Imperialism, the Highest Stage of Capitalism.* New York: International.

Levidow, Les, and Bob Young, eds. 1981. *Science, Technology, and the Labour Process: Marxist Studies.* London: CSE Books.

Levin, Sam. 2017. "New Facebook Tool Tells Users If They've Liked or Followed Russia's 'Troll Army.'" *Guardian,* December 22. https://www.theguardian.com/technology/2017/dec/22/facebook-tool-russia-troll-army-internet-research-agency/.

Levine, Yasha. 2014a. "Google Distances Itself from the Pentagon, Stays in Bed with Mercenaries and Intelligence Contractors." *Pando,* March 26. https://

pando.com/2014/03/26/google-distances-itself-from-the-pentagon-stays-in-bed-with-mercenaries-and-intelligence-contractors/.

Levine, Yasha. 2014b. "Oakland Emails Give Another Glimpse into the Google-Military-Surveillance Complex." *Pando,* March 7. https://pando.com/2014/03/07/the-google-military-surveillance-complex/.

Levine, Yasha. 2014c. "The Revolving Door between Google and the Department of Defense." *Pando,* April 23. https://pando.com/2014/04/23/the-revolving-door-between-google-and-the-department-of-defense/.

Levine, Yasha. 2018. *Surveillance Valley: The Secret Military History of the Internet.* New York: Hachette Books.

Levy, Steven. 1984. *Hackers: Heroes of the Computer Revolution.* New York: Doubleday.

Leyden, Joel. 2003. "Al-Qaeda: The 39 Principles of Holy War." Israel News Agency. September 4. http://www.israelnewsagency.com/Al-Qaeda.html.

Li, Minqui. 2009. *The Rise of China and the Demise of the Capitalist World Economy.* New York: Monthly Review Press.

Liang Qiao and Wang Xiangsui. 1999. *Unrestricted Warfare.* Beijing: PLA Literature and Arts.

Libicki, Martin. 2009. *Cyberdeterrence and Cyberwar.* Santa Monica, Calif.: RAND.

Lindsay, Jon. 2012. "International Cyberwar Treaty Would Quickly Be Hacked to Bits." *USA Today,* June 8. https://www.usnews.com/debate-club/should-there-be-an-international-treaty-on-cyberwarfare/international-cyberwar-treaty-would-quickly-be-hacked-to-bits.

Lindsay, Jon R. 2014. "The Impact of China on Cybersecurity: Fiction and Friction." *International Security* 39, no. 3: 7–47.

Lindsay, Jon R., Tai Ming Cheung, and Derek Reveron, eds. 2015. *China and Cybersecurity: Espionage, Strategy, and Politics in the Digital Domain.* Oxford: Oxford University Press.

Linebaugh, Peter, and Marcus Rediker. 2000. *The Many Headed Hydra: Sailors, Slaves, Commoners, and the Hidden History of the Revolutionary Atlantic.* Boston: Beacon Press.

Liu, Lydia H. 2010. *The Freudian Robot: Digital Media and the Future of the Unconscious.* Chicago: University of Chicago Press.

Loizos, Connie. 2017. "'When You Spend $100 Million on Social Media,' It Comes with Help, Says Trump Strategist." *TechCrunch,* November 8. https://techcrunch.com/2017/11/08/when-you-spend-100-million-on-social-media-it-comes-with-help-says-trump-strategist/.

Lovink, Geert. 2016. "On the Social Media Ideology." *e-flux* 75. http://www.e-flux.com/journal/75/67166/on-the-social-media-ideology/.

Maas, Peter. 2018. "Reality Winner Has Been in Jail for a Year. Her Prosecution Is Unfair and Unprecedented." *The Intercept* (blog), July 3. https://theintercept.com/2018/06/03/reality-winner-nsa-paul-manafort/.

Machkovech, Sam. 2018. "Twitter Begins Emailing the 677,775 Americans Who Took Russian Election Bait." *Ars Technica,* January 19. https://arstechnica.com/information-technology/2018/01/twitter-begins-emailing-the-677775-americans-who-took-russian-election-bait/.

Mackey, Robert. 2016. "With Facebook No Longer a Secret Weapon, Egypt's Protesters Turn to Signal." *Intercept,* April 26. https://theintercept.com/2016/04/26/facebook-no-longer-secret-weapon-egypts-protesters-turn-signal/.

Mackey, Robert. 2017. "We Knew Julian Assange Hated Clinton. We Didn't Know He Was Secretly Advising Trump." *Intercept,* November 15. https://theintercept.com/2017/11/15/wikileaks-julian-assange-donald-trump-jr-hillary-clinton/.

Mager, Astrid. 2012. "Algorithmic Ideology: How Capitalist Society Shapes Search Engines. Information." *Communication and Society* 15, no. 5: 769–87.

Mandel, Ernst. 1975. *Late Capitalism.* London: New Left Review.

Mao Zedong. (1938) 1967. "On Protracted War." In *Selected Works of Mao Tse-tung.* Vol. 2. Peking: Foreign Languages Press. http://www.marx2mao.com/Mao/PW38.html#s1/.

Maréchal, Nathalie. 2017. "Networked Authoritarianism and the Geopolitics of Information: Understanding Russian Internet Policy." *Media and Communication* 5, no. 1: 29–41.

Markoff, John. 2005. *What the Dormouse Said: How the Sixties Counterculture Shaped the Personal Computer Industry.* New York: Penguin.

Marx, Karl. 1973. *Grundrisse.* London: Penguin.

Marx, Karl. 1977. *Capital, Volume I.* Translated by Ben Fowkes. New York: Vintage.

Marx, Karl, and Friedrich Engels. (1848) 1964. *The Communist Manifesto.* Washington, DC: Washington Square Press.

Mason, Paul. 2012. *Why It's Kicking Off Everywhere: The New Global Revolutions.* London: Verso.

Massumi, Brian. 2015. *Ontopower: War, Powers, and the State of Perception.* Durham, N.C.: Duke University Press.

Maurer, Tim. 2018. *Cyber Mercenaries: The State, Hackers, and Power.* Cambridge: Cambridge University Press.

Maurer, Tim, and Scott Janz. 2014. "The Russia–Ukraine Conflict: Cyber and Information Warfare in a Regional Context." *ISN ETH Zurich,* October 17. https://www.files.ethz.ch/isn/187945/ISN_184345_en.pdf.

Mazzetti, Mark. 2014. *The Way of the Knife: The CIA, a Secret Army, and a War at the Ends of the Earth.* New York: Penguin.

Mazzucato, Mariana. 2013. *The Entrepreneurial State: Debunking Public vs. Private Sector Myths.* London: Anthem Press.

McCoy, Alfred. 2017. *In the Shadows of the American Century: The Rise and Decline of US Global Power.* Kindle ed. Chicago: Haymarket.

McCurry, Justin. 2017. "South Korea Spy Agency Admits Trying to Rig 2012 Presidential Election." *Guardian,* August 4. https://www.theguardian.com /world/2017/aug/04/south-koreas-spy-agency-admits-trying-rig-election -national-intelligence-service-2012/.

McLuhan, Marshall. 1994. *Understanding Media: The Extensions of Man.* Cambridge, Mass.: MIT Press.

Mennon, Rajan, and Eugene B. Rumer. 2015. *Conflict in Ukraine: The Unwinding of the Post–Cold War.* Cambridge, Mass.: MIT Press.

Metz, Cade. 2012. "Paul Baran, the Link between Nuclear War and the Internet." *Wired,* September 4. http://www.wired.co.uk/article/h-bomb-and-the -internet.

Metz, Steven, and James Kievit. 2013. *The Revolution in Military Affairs and Conflict Short of War.* https://ssi.armywarcollege.edu/pdffiles/PUB241.pdf.

Millar, Michael. 2011. "The Anti-Social Network: Avoiding Online Darkness." *BBC News,* April 11. http://www.bbc.com/news/business-13158351/.

Miller, Jacuques-Alain. 1990. "Microscopia: An Interview to the Reading of Television." In *Television: A Challenge to the Psychoanalytic Establishment,* by Jacques Lacan. Edited by Joan Copjec. Translated by Denis Hollier, Rosalind Krauss, Annette Michelson, and Jeffrey Mehlman. New York: W. W. Norton.

Miller, Robert, and Daniel Kuehl. 2009. *Cyberspace and the "First Battle" in 21st-Century War.* Washington, D.C.: Center for Technology and National Security Policy, National Defense University.

Mirowski, Philip. 2013. *Never Let a Serious Crisis Go to Waste: How Neoliberalism Survived the Financial Meltdown.* London: Verso.

Monereo, Manolo. 2017. "What's Next for Podemos?" *Jacobin,* May 4. https:// www.jacobinmag.com/2017/05/podemos-spain-congress-vistalegre-psoe -ciudananos.

Morozov, Evgeny. 2008. "An Army of Ones and Zeroes: How I Became a Soldier in the Georgia–Russia Cyberwar." *Slate,* August 5. http://www.slate.com /articles/technology/technology/2008/08/an_army_of_ones_and_ze roes.html.

Morozov, Evgeny. 2010. "In Defense of DDoS." *Slate,* December 13. http://www .slate.com/articles/technology/technology/2010/12/in_defense_of_ddos .html.

Morozov, Evgeny. 2012. "Why Hillary Clinton Should Join Anonymous." *Slate,* December 23. http://www.slate.com/articles/technology/future_tense /2012/04/internet_freedom_threat_posed_by_hillary_clinton_s_state_de partment_and_anonymous_.html.

Morozov, Evgeny. 2015. "Socialize the Data Centres!" *New Left Review* 91: 45–66. https://newleftreview.org/II/91/evgeny-morozov-socialize-the-data -centres.

Morozov, Evgeny. 2017. "Moral panic over fake news hides the real enemy—the digital giants." *Guardian,* January 8. https://www.theguardian.com/commentisfree/2017/jan/08/blaming-fake-news-not-the-answer-democracy-crisis.

Mosco, Vincent. 2017. *Becoming Digital: Toward a Post-Internet Society.* New York: Emerald.

Moss, Trefor. 2013. "Is Cyber War the New Cold War?" *Diplomat,* April 19. http://thediplomat.com/2013/04/is-cyber-war-the-new-cold-war/.

Moulier-Boutang, Yann. 2011. *Cognitive Capitalism.* Cambridge: Polity Press.

Müller-Maguhn, Andy, et al. 2014. "Map of the Stars: The NSA and GCHQ Campaign against German Satellite Companies." *Intercept,* September 14. https://theintercept.com/2014/09/14/nsa-stellar/.

Mundy, Liza. 2017. *Code Girls: The Untold Story of the American Women Code Breakers of World War II.* London: Hachette.

Murray, Williamson. 2012. *Hybrid Warfare: Fighting Complex Opponents from the Ancient World to the Present.* Cambridge: Cambridge University Press.

Nagle, Angela. 2017. *Kill All Normies: Online Culture Wars from 4Chan and Tumblr to Trump and the Alt-Right.* New York: Zero Books.

Nakashima, Ellen. 2010. "Dismantling of Saudi–CIA Web Site Illustrates Need for Clearer Cyberwar Policies." *Washington Post,* March 18. http://www.washingtonpost.com/wp-dyn/content/article/2010/03/18/AR2010031805464.html.

Nation, R. Craig. 1989. *War on War: Lenin, the Zimmerwald Left, and the Origins of the Communist International.* New York: Haymarket.

NATO Advanced Research Workshop. 2006. "Cyberwar–Netwar: Security in the Information Age." Paper presented at the NATO Advanced Research Workshop on Cyberwar–Netwar: Security in the Information Age, Lisbon.

Neocleous, Mark. 2014. *War Power, Police Power.* Edinburgh: Edinburgh University Press.

Neubauer, Merel. 2017. "Kurdish Movement 2.0: The Usage of Social Media Networks as a Political Space by a Nation without a State." https://pdflegend.com/download/kurdish-movement-2o-the-usage-of-social-media-networks-by-a-nation-without-a-state-_59f9a339d64ab23ff03de6d3_pdf.

Niva, Steve. 2013. "Disappearing Violence: JSOC and the Pentagon's New Cartography of Networked Warfare." *Security Dialogue* 44, no. 3: 185–202.

Noble, David. 1986. *Forces of Production: A Social History of Industrial Automation.* Oxford: Oxford University Press.

Noys, Benjamin, ed. 2011. *Communization and Its Discontents: Contestation, Critique and Contemporary Struggles.* New York: Autonomedia.

Noys, Benjamin. 2013. "War on Time: Occupy, Communization and the Military Question." https://libcom.org/library/war-time-occupy-communization-military-question-benjamin-noys.

Nunes, Rodrigo. 2014a. "Notes toward a Rethinking of the Militant." In *Communism in the 21st Century*, edited by Shannon Brincat, 3:163–87. Santa Barbara, Calif.: ABC-Clio.

Nunes, Rodrigo. 2014b. *Organisation of the Organisationless: Collective Action after Networks*. London: Mute and Post-Media Lab.

O'Donovan, Caroline. 2018. "Employees of Another Major Tech Company Are Petitioning Government Contracts." *BuzzFeed News*, June 26. https://www .buzzfeednews.com/article/carolineodonovan/salesforce-employees-push -back-against-company-contract#.podOEwrAj.

Olson, Parmy. 2012. "Amid ACTA Outcy [sic], Politicians Don Anonymous Guy Fawkes Masks." *Forbes*, January 27. https://www.forbes.com/sites /parmyolson/2012/01/27/amid-acta-outcy-politicians-don-anonymous -guy-fawkes-masks/#4a49f9c65064/.

O'Neill, Cathy. 2016. *Weapons of Math Destruction: How Big Data Increases Inequality and Threatens Democracy*. New York: Random House.

O'Reilly, Tim. 2005. "What Is Web 2.0 Design Patterns and Business Models for the Next Generation of Software." *O'Reilly* (blog), September 30. https:// www.oreilly.com/pub/a/web2/archive/what-is-web-20.html.

O'Rourke, Lindsey. 2016. "The U.S. Tried to Change Other Countries' Governments 72 Times during the Cold War." *Washington Post*, December 23. https://www.washingtonpost.com/news/monkey-cage/wp/2016/12/23 /the-cia-says-russia-hacked-the-u-s-election-here-are-6-things-to-learn-from -cold-war-attempts-to-change-regimes/.

Paasonen, Susanna. 2011. *Carnal Resonance Affect and Online Pornography*. Cambridge, Mass.: MIT Press.

Paganini, Pierlugui. 2016. "Phineas Fisher Hacked a Bank to Support Anticapitalists in the Rojava Region." *Security Affairs*, May 20. http://security affairs.co/wordpress/47496/cyber-crime/phineas-fisher-hacked-bank.html.

Parakilas, Sandy. 2018. "Facebook Wants to Fix Itself. Here's a Better Solution." *Wired*, January 30. https://www.wired.com/story/facebook-wants-to-fix -itself-heres-a-better-solution/.

Pariser, Eli. 2011. *The Filter Bubble: How the New Personalized Web Is Changing What We Read and How We Think*. New York: Penguin.

Pasquale, Frank. 2015. *Black Box Society: The Secret Algorithms That Control Money and Information*. Cambridge, Mass.: Harvard University Press.

Patrikarakos, David. 2017. *War in 140 Characters: How Social Media Is Reshaping Conflict in the Twenty-First Century*. New York: Basic Books.

Peirano, Marta. 2018. "Built-In Values: On the Politicization of Media Platforms." Panel at Transmediale, Berlin, Germany, March 22. https://www.youtube .com/watch?v=VxHKWZ2pNno.

Peters, Benjamin. 2016. *How Not to Network a Nation: The Uneasy History of the Soviet Internet*. Cambridge, Mass.: MIT Press.

Pietrzyk, Kamilla. 2010. "Activism in the Fast Lane: Social Movements and the

Neglect of Time." *Fast Capitalism* 1. http://www.uta.edu/huma/agger /fastcapitalism/7_1/pietrzyk7_1.html.

Pichai, Sundar. 2018. "AI at Google: Our Principles." *Google* (blog), June 7. https:// www.blog.google/technology/ai/ai-principles/.

Pirani, Simon. 2010. *Change in Putin's Russia: Power, Money and People.* London: Pluto.

Plotke, David. 2012. "Occupy Wall Street, Flash Movements, and American Politics." *Dissent Magazine,* August 15.

Porche, Isaac R., Christopher Paul, Michael York, Chad C. Serena, Jerry M. Sollinger, Elliot Axelband, Endy M. Daehner, and Bruce J. Held. 2013. *Redefining Information Warfare Boundaries for an Army in a Wireless World.* Santa Monica, Calif.: RAND.

Powers, Shawn, and Michael Jablonski. 2015. *The Real Cyber War: The Political Economy of Internet Freedom.* Kindle ed. Chicago: University of Chicago Press.

Pozen, David E. 2005. "The Mosaic Theory, National Security, and the Freedom of Information Act." *Yale Law Journal* 115, no. 628: 630–78.

Pufeng, Wang. 1995. "The Challenge of Information Warfare, China Military Science." https://fas.org/irp/world/china/docs/iw_mg_wang.htm.

Qiu, Jack Linchuan. 2009. *Working-Class Network Society: Communication Technology and the Information Have-Less in Urban China.* Cambridge, Mass.: MIT Press.

Radio Netherlands. 2013. "Twitter's Activist Roots: How Twitter's Past Shapes Its Use as a Protest Tool." November 15. https://www.pressenza.com/2013/11 /twitters-activist-roots-twitters-past-shapes-use-protest-tool/.

Radloff, Jennifer, and Grady Johnson. 2012. "Collateral Damage of the Cyberwar in Syria." GenderIT. October 24. http://www.genderit.org/es/node/3678.

RAND. 2008. "Paul Baran and the Origins of the Internet." https://www.rand .org/about/history/baran.html.

Rantapelkonen, Jari, and Mirva Salminen, eds. 2013. *The Fog of Cyberwar.* Helsinki: Finnish National Defence University.

Reglado, Daniel, Nart Villeneuve, and John Scott-Railton. 2015. "Behind the Syrian Conflict's Digital Front Lines." Fireye Threat Intelligence. http:// www.fireeye.com/current-threats/threat-intelligence-reports.html.

Retort. 2005. *Afflicted Powers: Capital and Spectacle in a New Age of War.* New York: Verso.

Reynolds, Matthew. 2017. "This Is What You Need to Know about Those Russian Facebook Ads." *Wired,* November 2. http://www.wired.co.uk/article /facebook-twitter-russia-congress-fake-ads-2016-election-trump/.

Rid, Thomas. 2013. *Cyber War Will Not Take Place.* Oxford: Oxford University Press.

Rid, Thomas. 2016. *Rise of the Machines: A Cybernetic History.* New York: W. W. Norton.

Rid, Thomas, and Ben Buchanan. 2015. "Attributing Cyber-Attacks." *Journal*

of Strategic Studies 38, no. 1/2: 4–37. https://doi.org/10.1080/01402390.20
14.977382.

Rid, Thomas, and Peter McBurney. 2012. "Cyber-Weapons." *RUSI Journal* 157,
no. 1: 6–13.

Ridgeway, Renee. 2017. "Who's Hacking Whom?" *Limn* 8. http://limn.it/the
-spy-who-pwned-me/.

Risen, James, and Laura Poitras. 2013. "NSA Report Outlined Goals for More
Power." *New York Times*, November 23. http://www.nytimes.com/2013/11/23
/us/politics/nsa-report-outlined-goals-for-more-power.html.

Ritchie, Kevin. 2016. "Hack the Job Market with This Growing Number of
Post-secondary Cyber-security Courses." *NowToronto*, October 19. https://
nowtoronto.com/lifestyle/class-action/cyber-security-courses-hack-the
-job-market/.

Robertson, Jordan Michael Riley, and Andrew Willis. 2016. "How to Hack an
Election." *Bloomberg Businessweek*, March 31. https://www.bloomberg.com
/features/2016-how-to-hack-an-election/.

Robins, Kevin, and Frank Webster. 1988. "Cybernetic Capitalism: Information, Technology, Everyday Life." In *The Political Economy of Information*,
edited by Vincent Mosco and Janet Wasko, 44–75. Madison: University of
Wisconsin Press.

Rodley, Chris. 2104. "When Memes Go to War: Viral Propaganda in the 2014
Gaza-Israel Conflict." *Fibreculture Journal* 27. http://twentyseven.fibrecul
turejournal.org/2016/03/18/fcj-200-when-memes-go-to-war-viral-propa
ganda-in-the-2014-gaza-israel-conflict/.

Romanyshyn, Yuliana. 2018a. "Goodbye, ATO: Ukraine Officially Changes
Name of Donbas War." *Kyiv Post*, February 20. https://www.kyivpost
.com/ukraine-politics/goodbye-ato-hello-taking-measures-ensure-national
-security-defense-repulsing-deterring-armed-aggression-russian-federation
-donetsk-luhansk-oblasts.html.

Romanyshyn, Yuliana. 2018b. "Investigation Unveils Poroshenko's Secret Vaca-
tion in Maldives." *Kyiv Post*, January 19. https://www.kyivpost.com/ukraine
-politics/investigation-unveils-poroshenkos-secret-vacation-maldives.html.

Ronell, Avital. 2002. *Stupidity*. Chicago: University of Chicago Press.

Ronfeldt, David, John Arquilla, Graham E. Fuller, and Melissa Fuller. 1998. *The
Zapatista Social Netwar in Mexico*. Santa Monica, Calif.: RAND.

Rosenberg, Scott. 2017. "Firewalls Don't Stop Hackers. AI Might." *Wired*, September 27. https://www.wired.com/story/firewalls-dont-stop-hackers
-ai-might/.

Rosenzweig, Paul. 2012. *Cyber Warfare: How Conflicts in Cyberspace Are Challenging
America and Changing the World*. Santa Barbara, Calif.: Praeger.

Rossiter, Ned. 2006. *Organized Networks: Media Theory, Creative Labour, New
Institutions*. New York: naio10.

Sadri, Houman A., and Nathan Burns. 2010. "The Georgia Crisis: A New Cold

War on the Horizon?" *Caucasian Review of International Affairs* 4, no. 2: 122–44.

Sageman, Marc. 2008. *Leaderless Jihad: Terror Networks in the Twenty-First Century.* Philadelphia: University of Pennsylvania Press.

Sakwa, Richard. 2013. "The Cold Peace: Russo-Western Relations as a Mimetic Cold War." *Cambridge Review of International Affairs* 26, no. 1: 203–24.

Sakwa, Richard. 2014. *Putin Redux.* New York: Routledge.

Sampson, Tony D. 2012. *Virality: Contagion Theory in the Age of Networks.* Minneapolis: University of Minnesota Press.

Sanger, David E. 2018. *The Perfect Weapon: War, Sabotage, and Fear in the Cyber Age.* New York: Crown.

Sauter, Molly. 2014. *The Coming Swarm: DDOS Actions, Hacktivism, and Civil Disobedience on the Internet.* London: Bloomsbury.

Scahill, Jeremy. 2013. *Dirty Wars: The World Is a Battlefield.* New York: Nation Books.

Scahill, Jeremy. 2016. *The Assassination Complex.* New York: Simon and Schuster.

Scarry, Elaine. 2104. *Thermonuclear Monarchy: Choosing between Democracy and Doom.* New York: W. W. Norton.

Shachtman, Noah, and Robert Beckhusen. 2012. "Hamas Shoots Rockets at Tel Aviv, Tweeting Every Barrage." *Wired,* November 15. http://wired.com /2012/11/gaza-social-media-war/.

Scharre, Paul. 2018. *Army of None: Autonomous Weapons and the Future of War.* New York: W. W. Norton.

Scheer, Robert, and Narda Zacchino. 1983. *With Enough Shovels: Reagan, Bush, and Nuclear War.* New York: Martin Secker and Warburg.

Schiller, Daniel. 1999. *Digital Capitalism: Networking the Global Market System.* Cambridge, Mass.: MIT Press.

Schiller, Dan. 2014. *Digital Depression: Information Technology and Economic Crisis.* Urbana: University of Illinois Press

Schneier, Bruce. 2010. "The Story behind the Stuxnet Virus." *Forbes,* October 7. https://www.forbes.com/2010/10/06/iran-nuclear-computer-technology -security-stuxnet-worm.html#195a8bef51e8/.

Schneier, Bruce. 2013. "US Offensive Cyberwar Policy." *Schneier on Security* (blog), June 21. https://www.schneier.com/blog/archives/2013/06/us_offensive _cy.html.

Schneier, Bruce. 2015. *Data and Goliath: The Hidden Battles to Collect Your Data and Control Your World.* London: W. W. Norton.

Schneier, Bruce. 2016. "Someone Is Learning How to Take Down the Internet." *Schneier on Security* (blog), September 13. https://www.schneier.com/blog /archives/2016/09/someone_is_lear.html.

Schneier, Bruce. 2018. *Click Here to Kill Everybody: Security and Survival in a Hyperconnected World.* New York: Naughton.

Schrage, Michael. 2013. "Businesses Are Now Combatants in a Cyberwar

with China and Iran." *Harvard Business Review,* February 6. https://hbr
.org/2013/02/businesses-are-now-combatants/.

Schumpeter, Joseph A. 1942. *Capitalism, Socialism and Democracy.* London:
Routledge.

Scott-Railton, John. 2013. "Revolutionary Risks: Cyber Technology and Threats
in the 2011 Libyan Revolution." Center on Irregular Warfare and Armed
Groups, U.S. Naval War College, Newport, R.I. http://www.usnwc.edu
/ciwag/.

Scott-Railton, John. 2018. "Fit Leaking: When a Fitbit Blows Your Cover."
Citizen Lab. January 28. http://www.johnscottrailton.com/fit-leaking/.

Segal, Adam. 2016. *The Hacked World Order: How Nations Fight, Trade, Maneuver
and Manipulate in the Digital Age.* New York: Public Affairs.

Segal, Adam. 2017. "How China Is Preparing for Cyberwar." *Christian Science Monitor,* March 20. https://www.csmonitor.com/World/Passcode
/Passcode-Voices/2017/0320/How-China-is-preparing-for-cyberwar/.

Sergatskova, Katerina. 2016. "Has Ukraine become More Dangerous for Journalists than Russia?" *Moscow Times,* July 27. https://themoscowtimes.com
/articles/has-ukraine-become-more-dangerous-for-journalists-than-russia
-54741.

Shachtman, Noah. 2009. "Wage Cyberwar against Hamas, Surrender Your
PC." *Wired,* August 1. https://www.wired.com/2009/01/israel-dns-hack/.

Shane, Scott. 1995. *Dismantling Utopia: How Information Ended the Soviet Union.*
New York: Ivan R. Dee.

Sharma, Sarah. 2013. "Critical Time." *Communication and Critical/Cultural Studies* 10, no. 2–3: 312–18.

Shaw, Ian. 2016. *Predator Empire: Drone Warfare and Full Spectrum Dominance.*
Minneapolis: University of Minnesota Press.

Shilling, Erk. 2017. "From 'Byzantine Hades' to 'Titan Rain,' Cyber Attack
Code Names Are Sci-Fi Poetry." *Atlas Obscura,* February 17. http://www
.atlasobscura.com/articles/from-byzantine-hades-to-titan-rain-cyber-attack
-code-names-are-scifi-poetry.

Silver, Beverly. 2003. *Forces of Labor: Workers' Movements and Globalization since
1870.* Cambridge: Cambridge University Press.

Simonite, Tom. 2012. "Stuxnet Tricks Copied by Computer Criminals." *MIT
Technology Review,* September 19. https://www.technologyreview.com
/s/429173/stuxnet-tricks-copied-by-computer-criminals/.

Sindelar, Daisy. 2014. "The Kremlin's Troll Army." *Atlantic,* August 12. http://
www.theatlantic.com/international/archive/2014/08/the-kremlins-troll
-army/375932/.

Singal, Jesse. 2016. "Why Did WikiLeaks Help Dox Most of Turkey's Adult
Female Population?" *Select/All,* July 27. http://nymag.com/selectall/2016/07
/why-did-wikileaks-help-dox-most-of-turkeys-adult-female-population
.html.

Singer, P. W. 2009. *Wired for War: The Robotics Revolution and Conflict in the 21st Century.* New York: Penguin.

Singer, P. W., and August Cole. 2015. *Ghost Fleet: A Novel of the Next World War.* New York: Houghton Mifflin.

Singer, P. W., and Allan Friedman. 2014. *Cybersecurity and Cyberwar: What Everyone Needs to Know.* Oxford: Oxford University Press.

Sjoberg, Laura. 2014. *Gender, War and Conflict.* Cambridge: Polity.

Slaughter, Anne-Marie, and Elizabeth Weingarten. 2016. "The National Security Issue No One Is Talking About." *Time,* April 16. http://time.com/4290563 /women-in-cybersecurity/.

Sloan, Elinor C. 2002. *The Revolution in Military Affairs.* Montreal: McGill University Press.

Sloterdijk, Peter. 2011. *Bubbles: Spheres.* Vol. 1. *Microspherology.* Los Angeles, Calif.: Semiotext(e).

Sly, Liz. 2018. "U.S. Soldiers Are Revealing Sensitive and Dangerous Information by Jogging." *Washington Post,* January 29. https://www.washingtonpost .com/world/a-map-showing-the-users-of-fitness-devices-lets-the-world -see-where-us-soldiers-are-and-what-they-are-doing/2018/01/28/86915662 -0441-11e8-aa61-f3391373867e_story.html.

Smith, Brad. 2017. "The Need for Urgent Collective Action to Keep People Safe Online: Lessons from Last Week's Cyberattack." *Microsoft on the Issues* (blog), May 14. https://blogs.microsoft.com/on-the-issues/2017/05/14 /need-urgent-collective-action-keep-people-safe-online-lessons-last-weeks -cyberattack/.

Soldatov, Andrei, and Irina Borogan. 2011. *The New Nobility: The Restoration of Russia's Security State and the Enduring Legacy of the KGB.* Philadelphia: PublicAffairs.

Soldatov, Andrei, and Irina Borogan. 2015. *The Red Web: The Struggle between Russia's Digital Dictators and the New Online Revolutionaries.* New York: Public Affairs.

Soshnikov, Andrei. 2017. "Inside a Pro-Russia Propaganda Machine in Ukraine." *BBC Russian,* November 13. http://www.bbc.com/news/blogs-trending -41915295/.

Spaiser, Viktoria, Thomas Chadefaux, Karsten Donnay, Fabian Russmann, and Dirk Helbing. 2017. "Communication Power Struggles on Social Media: A Case Study of the 2011–12 Russian Protests." *Journal of Information Technology and Politics* 14, no. 2: 132–53.

Spufford, Francis. 2010. *Red Plenty.* London: Faber and Faber.

Srnicek, Nick. 2016. *Platform Capitalism.* London: Verso.

Stamos, Alex. 2017. "An Update on Information Operations on Facebook." Facebook. September 6. https://newsroom.fb.com/news/2017/09/information -operations-update/.

Steinmetz, Kevin E. 2016. *Hacked: A Radical Approach to Hacker Culture and Crime.* New York: New York University Press.

Sterling, Bruce. 2007. "Here Comes the Cyber Cold War, Spies Declare Eagerly." *Wired,* December 11. https://www.wired.com/2007/12/here-comes-th-1/.

Stiennon, Richard. 2015. *There Will Be Cyberwar: How the Move to Network-Centric War Fighting Has Set the Stage for Cyberwar.* New York: Harvester.

Streitfeld, David. 2017. "'The Internet Is Broken': @ev Is Trying to Salvage It." *New York Times,* May 20. https://www.nytimes.com/2017/05/20/technology/evan-williams-medium-twitter-internet.html.

Subramanian, Samanth. 2017. "Inside the Macedonian: Fake-News Complex." *Wired,* February 15. https://www.wired.com/2017/02/veles-macedonia-fake-news/.

Suciu, Peter. 2014. "Why Cyber Warfare Is So Attractive to Small Nations." *Fortune,* December 21. http://fortune.com/2014/12/21/why-cyber-warfare-is-so-attractive-to-small-nations/.

Szwed, Robert. 2016. "Framing of the Ukraine–Russia Conflict in Online and Social Media." NATO Strategic Communication Centre of Excellence. https://www.stratcomcoe.org/framing-ukraine-russia-conflict-online-and-social-media.

Tactical Technology and Frontline Defenders. 2009. "Security in a Box: Digital Security Tools and Tactics." https://securityinabox.org/en/.

Talley, Sue. 2013. "In Cyber Warfare, Education Is Our Most Powerful Weapon." *Huffington Post,* February 11. https://www.huffingtonpost.com/sue-talley-edd/in-cyber-warfare-education_b_2244950.html.

Taplin, Jonathan. 2017. *Move Fast and Break Things: How Facebook, Google, and Amazon Cornered Culture and Undermined Democracy.* New York: Little, Brown.

Tarnoff, Ben. 2018. "Can Silicon Valley Workers Rein in Big Tech from Within?" *Guardian,* August 9. https://www.theguardian.com/commentisfree/2018/aug/09/silicon-valley-tech-workers-labor-activism.

Taubman, Philip. 2003. *Secret Empire: Eisenhower, the CIA, and the Hidden Story of America's Space Espionage.* New York: Simon and Schuster.

Taylor, Samuel. 2012. "Blockades, Social Media and War Crimes: From Cast-Lead to the Mavi Marmara." *Eye of the Needle,* June 13. https://throughtheneedleseye.wordpress.com/2012/06/13/blockades-social-media-war-crimes-from-cast-lead-to-the-mavi-marmara/.

Teschke, Bruno. 2008. "Marxism." In *The Oxford Handbook of International Relations,* edited by Christian Reus-Smith and Duncan Snidal, 163–87. Oxford: Oxford University Press.

Thistlethwaite, Susan Brooks. 2016. "The Cyber War on Women: Anita Sarkeesian Cancels Utah Lecture after Shooting Threats." *Huffington Post,* October

16. https://www.huffingtonpost.com/rev-dr-susan-brooks-thistlethwaite/the-cyber-war-on-women-an_b_5995154.html.

Thomas, Timothy. 2000. "The Russian View of Information War." In *The Russian Armed Forces at the Dawn of the Millennium,* edited by Michael Cruchter, 335–60. Carlisle, Pa.: U.S. Army War College. http://www.dtic.mil/dtic/tr/fulltext/u2/a423593.pdf.

Thomas-Noone, Brendan. 2017. "Cyberwar and War in Space: Making SSBNs More Dangerous." *Interpreter,* November 26. https://www.lowyinstitute.org/the-interpreter/cyberwar-and-war-space-making-ssbns-more-dangerous.

Tikk, Eneken, Kadri Kaska, and Liis Vihul. 2010. *International Cyber Incidents: Legal Considerations.* Tallinn, Estonia: Cooperative Cyber Defence Centre of Excellence.

Timberg, Craig. 2017. "In Trump's America, Black Lives Matter Activists Grow Wary of Their Smartphones." *Washington Post,* June 1. https://www.washingtonpost.com/business/technology/fearing-surveillance-in-the-age-of-trump-activists-study-up-on-digital-anonymity/2017/05/20/186e8ba0-359d-11e7-b4ee-434b6d506b37_story.html.

Tiqqun. 2001. "The Cybernetic Hypothesis." Vol. 2. https://theanarchistlibrary.org/library/tiqqun-the-cybernetic-hypothesis.

Toal, Gerrard. 2017. *Near Abroad: Putin, the West, and the Contest over Ukraine and the Caucuses.* Oxford: Oxford University Press.

Toler, Aric. 2014. "Ukrainian Hackers Leak Russian Interior Ministry Docs with 'Evidence' of Russian Invasion." *Global Voices,* December 13. http://globalvoicesonline.org/2014/12/13/ukrainian-hackers-leak-russian-interior-ministry-docs-with-evidence-of-russian-invasion/.

Tomšič, Samo. 2015. *The Capitalist Unconscious: Marx and Lacan.* London: Verso.

Townsend, Mark, Paul Harris, Alex Duval Smith, Dan Sabbagh, and Josh Halliday. 2010. "WikiLeaks Backlash: The First Global Cyber War Has Begun, Claim Hackers." *Guardian,* December 11. https://www.theguardian.com/media/2010/dec/11/wikileaks-backlash-cyber-war.

Trotsky, Leon. 1932. "In Defence of October: A Speech Delivered in Copenhagen, Denmark in November 1932." Marxists Internet Archive. https://www.marxists.org/archive/trotsky/1932/11/oct.htm.

Trotsky, Leon. (1937) 1973. *The Revolution Betrayed.* London: Pathfinder.

Tucker, Patrick. 2017. "For the US Army, 'Cyber War' Is Quickly Becoming Just 'War.'" *Defense One,* February 9. http://www.defenseone.com/technology/2017/02/us-army-cyber-war-quickly-becoming-just-war/135314/?oref=search_cyberwar.

Tufekci, Zeynep. 2016. "WikiLeaks Put Women in Turkey in Danger, for No Reason." *Huffington Post,* July 27. https://www.huffingtonpost.com/zeynep-tufekci/wikileaks-erdogan-emails_b_11158792.html.

Tufekci, Zeynep. 2017. *Twitter and Tear Gas: The Power and Fragility of Networked Protest*. New Haven, Conn.: Yale University Press.

Turner, Fred. 2006. *From Counterculture to Cyberculture: Stewart Brand, the Whole Earth Network, and the Rise of Digital Utopianism*. Chicago: University of Chicago Press.

Turse, Nick. 2012. *The Changing Face of Empire: Special Ops, Drones, Spies, Proxy Fighters, Secret Bases and Cyberwarfare*. Chicago: Haymarket Books.

Tynes, Robert. 2017. "When Ghost Sec Goes Hunting." *Limn* 8. http://limn.it/when-ghostsec-goes-hunting/.

U.S. Army Command and General Staff College. 2014. *Inevitable Evolutions: Punctuated Equilibrium and the Revolution in Military Affairs*. New York: U.S. Army Command and General Staff College.

U.S. Computer Emergency Team. 2017. "Alert (TA17–164A) Hidden Cobra: Korea's DDoS Botnet Infrastructure." June 13. https://www.us-cert.gov/ncas/alerts/TA17-164A.

U.S. Department of Justice. 2018a. "Grand Jury Indicts Thirteen Russian Individuals and Three Russian Companies for Scheme to Interfere in the United States Political System." *Justice News*, February 16. https://www.justice.gov/opa/pr/grand-jury-indicts-thirteen-russian-individuals-and-three-russian-companies-scheme-interfere.

U.S. Department of Justice. 2018b. "Grand Jury Indicts 12 Russian Intelligence Officers for Hacking Offenses Related to the 2016 Election." *Justice News*, July 13. https://www.justice.gov/opa/pr/grand-jury-indicts-12-russian-intelligence-officers-hacking-offenses-related-2016-election.

Üstündağ, Nazan. 2016. "Self-Defense as a Revolutionary Practice in Rojava, or How to Unmake the State." *South Atlantic Quarterly* 115, no. 1: 197–2010.

Valeriano, Brandon, and Ryan Maness. 2012. "The Fog of Cyberwar." *Foreign Affairs*, November 21. https://www.foreignaffairs.com/articles/2012-11-21/fog-cyberwar.

Valeriano, Brandon, and Ryan Maness. 2015. *Cyber War versus Cyber Realities: Cyber Conflict in the International System*. Oxford: Oxford University Press.

van Mensvoort, Koert. 2012. "Internet Traffic Is Now 51% Non-Human." *Next Nature Network* (blog), March 18. https://www.nextnature.net/2012/03/internet-traffic-is-now-51-non-human/.

Ventre, Daniel, ed. 2011. *Cyberwar and Information Warfare*. Hoboken, N.J.: John Wiley.

Virilio, Paul. (1977) 1986. *Speed and Politics: An Essay on Dromology*. New York: Semiotext(e).

Virilio, Paul. (1978) 1990. *Popular Defense and Ecological Struggles*. New York: Semiotext(e).

Virilio, Paul. 2000. *The Information Bomb*. London: Verso.

Virilio, Paul. 2002. *Crepuscular Dawn*. New York: Semiotext(e).

Virilio, Paul, and Sylvère Lotringer. (1983) 1997. *Pure War*. Rev. ed. New York: Semiotext(e).

Virilio, Paul, and Phillipe Petit. 1996. *Politics of the Very Worst*. New York: Semiotext(e).

Voelz, Glenn J. 2015. *The Rise of iWar: Identity, Information and the Individualization of Modern Warfare*. Carlisle, Pa.: U.S. Army War College, Strategic Studies Institute.

Volz, Dustin. 2018. "Trump, Seeking to Relax Rules on U.S. Cyberattacks, Reverses Obama Directive." *Wall Street Journal*, August 15. https://www.wsj .com/articles/trump-seeking-to-relax-rules-on-u-s-cyberattacks-reverses -obama-directive-1534378721.

Walker, Shaun, and Oksana Grytsenko. 2014. "Text Messages Warn Ukraine Protesters They Are 'Participants in Mass Riot.'" *Guardian*, January 21. https://www.theguardian.com/world/2014/jan/21/ukraine-unrest-text -messages-protesters-mass-riot.

Ward, Mark. 2017. "Smart Machines v Hackers: How Cyber Warfare Is Escalating." *BBC News*, March 10. http://www.bbc.com/news/business-38403426/.

Wark, McKenzie. 2004. *A Hacker Manifesto*. Cambridge, Mass.: Harvard University Press.

Wark, McKenzie. 2014. "William Gibson's The Peripheral." *Public Seminar*, November 2. http://www.publicseminar.org/2014/11/william-gibsons -the-peripheral/.

Weaver, Nicolas. 2013. "Our Government Has Weaponized the Internet. Here's How They Did It." *Wired*, November 13. https://www.wired.com/2013/11 /this-is-how-the-internet-backbone-has-been-turned-into-a-weapon/.

Weedon, Jen, William Nuland, and Alex Stamos. 2017. "Information Operations and Facebook: Version 1.0." *Facebook News Room* (blog), April 27. https:// fbnewsroomus.files.wordpress.com/2017/04/facebook-and-information -operations-v1.pdf.

Weimann, Gabriel. 2006. *Terror on the Internet: The New Arena, the New Challenges*. Washington, D.C.: U.S. Institute of Peace.

Weizman, Eyal. 2012. *Hollow Land: Israel's Architecture of Occupation*. New York: Verso.

White, Jeremy. 2017. "Facebook under Fire as It Reveals 150m Americans Saw Russian Election Propaganda." *Independent*, November 1. https://www .independent.co.uk/news/world/americas/us-politics/facebook-russia -ads-trump-election-2016-how-many-americans-saw-a8031881.html.

Wilson, Andrew. 2014. *Ukraine Crisis: What It Means for the West*. New Haven, Conn.: Yale University Press.

Wilson, Ash. 2017. "Infographic: Universities Still Struggle to Provide Cybersecurity Education." *Cloud Security* (blog), August 22. https://blog.cloudpassage .com/2017/08/22/universities-still-struggle-cybersecurity-education/.

Wilson, Louise. 1994. "Cyberwar, God and Television: Interview with Paul Virilio." *CTheory*, January 12. http://www.ctheory.net/articles.aspx ?id=62.

Winthrop-Young, Geoffrey. 2011. *Kittler and the Media*. Cambridge: Polity.

Wolfson, Todd. 2014. *Digital Rebellion: The Birth of the Cyber Left*. Urbana: University of Illinois Press.

Wood, Blake. 2005. "Computers, Codes, and Devices: The U.S. Nuclear Weapons Design Experience." Presented to the Project on Nuclear Issues fall conference, September 28–29.

Wood, Leslie J. 2014. *Crisis and Control: The Militarization of Protest Policing*. London: Pluto Press.

Woodman, Spencer. 2016. "Documents Suggest Palantir Could Help Power Trump's 'Extreme Vetting' of Immigrants." *Verge*, December 21. https://www.theverge.com/2016/12/21/14012534/palantir-peter-thiel-trump-immigrant-extreme-vetting.

Worth, Owen. 2005. *Hegemony, International Political Economy and Post-Communist Russia*. Aldershot, U.K.: Ashgate.

Worth, Owen. 2015. *Rethinking Hegemony*. London: Palgrave.

Wu, Susan. 2017. "It Is Time for Innovators to Take Responsibility for Their Creations." *Wired*, January 25. https://www.wired.com/story/its-time-for-innovators-to-take-responsibility-for-their-creations/.

Wu, Tim. 2017. *The Attention Merchants: The Epic Scramble to Get Inside Our Heads*. New York: Vintage.

Yonah, Jeremy. 2018. "Artificial Intelligence Cyber-hacking Arms Race at Full Throttle." *Jerusalem Post Israel News*, January 26. http://www.jpost.com/Israel-News/Artificial-intelligence-cyber-hacking-arms-race-at-full-throttle-539886/.

Yong-Soo, Eun, and Judith Sita Aßmann. 2016. "Cyberwar: Taking Stock of Security and Warfare in the Digital Age." *International Studies Perspectives* 17, no. 3: 343–60. https://doi.org/10.1111/insp.12073.

Young, Peter, and Rosemary Bennett. 2017. "University Secrets Are Stolen by Cybergangs: Scientific Research Targeted by Hackers." *Times*, September 5. https://www.thetimes.co.uk/article/university-secrets-are-stolen-by-cybergangs-oxford-warwick-and-university-college-london-rozsmf56z.

Zetter, Kim. 2014a. *Countdown to Zero Day: Stuxnet and the Launch of the World's First Digital Weapon*. New York: Crown.

Zetter, Kim. 2014b. "Meet MonsterMind, the NSA Bot That Could Wage Cyberwar Autonomously." *Wired*, August 13. https://www.wired.com/2014/08/nsa-monstermind-cyberwarfare/.

Zetter, Kim. 2016a. "Apple's FBI Battle Is Complicated. Here's What's Really Going On." *Wired*, February 18. https://www.wired.com/2016/02/apples-fbi-battle-is-complicated-heres-whats-really-going-on/.

Zetter, Kim. 2016b. "Inside the Cunning, Unprecedented Hack of Ukraine's Power Grid." *Wired,* March 3. https://www.wired.com/2016/03/inside-cunning-unprecedented-hack-ukraines-power-grid/.

Žižek, Slavoj. 1989. *The Sublime Object of Ideology.* London: Verso.

Žižek, Slavoj. 1997a. "Multiculturalism; or, The Cultural Logic of Multinational Capitalism." *New Left Review* 1, no. 125. https://newleftreview.org/I/225/slavoj-zizek-multiculturalism-or-the-cultural-logic-of-multinational-capitalism/.

Žižek, Slavoj. 1997b. *The Plague of Fantasies.* London: Verso.

Žižek, Slavoj. 1999. *The Ticklish Subject: The Absent Centre of Political Ontology.* London: Verso.

Žižek, Slavoj. N.d. "With or Without Passion: What's Wrong with Fundamentalism." Part I. http://www.lacan.com/zizpassion.htm.

Миротворец. N.d. "Information for Law Enforcement Authorities and Special Services about Pro-Russian Terrorists, Separatists, Mercenaries, War Criminals, and Murderers." Center for Research of Signs of Crimes against the National Security of Ukraine, Peace, Humanity, and the International Law. https://myrotvorets.center/.

Index

211

Nick Dyer-Witheford is associate professor of information and media studies at the University of Western Ontario. He is author of *Cyber-Marx: Cycles and Circuits of Struggle in High Technology Capitalism* and *Cyber-Proletariat: Global Labour in the Digital Vortex* and coauthor of *Digital Play: The Interaction of Technology, Culture, and Marketing* and *Games of Empire: Global Capitalism and Video Games* (Minnesota, 2009).

Svitlana Matviyenko is assistant professor of communication at Simon Fraser University and coeditor of *The Imaginary App* and *Lacan and the Posthuman*.